T0360828

Metabolomics and Microbiomics

Metabolomics and Microbiomics

Personalized Medicine from the Fetus to the Adult

Vassilios Fanos

Post-Graduate School of Pediatrics
University of Cagliari
Cagliari, Italy

Neonatal Intensive Care Unit
Neonatal Pathology and Neonatal Section
University of Cagliari
Cagliari, Italy

AMSTERDAM • BOSTON • HEIDELBERG • LONDON
NEW YORK • OXFORD • PARIS • SAN DIEGO
SAN FRANCISCO • SINGAPORE • SYDNEY • TOKYO
Academic Press is an Imprint of Elsevier

Academic Press is an imprint of Elsevier
125 London Wall, London EC2Y 5AS, United Kingdom
525 B Street, Suite 1800, San Diego, CA 92101-4495, United States
50 Hampshire Street, 5th Floor, Cambridge, MA 02139, United States
The Boulevard, Langford Lane, Kidlington, Oxford OX5 1GB, United Kingdom

Library of Congress Cataloging-in-Publication Data
A catalog record for this book is available from the Library of Congress

British Library Cataloguing-in-Publication Data
A catalogue record for this book is available from the British Library

ISBN: 978-0-12-805305-8

For information on all Academic Press publications
visit our website at http://www.elsevier.com/

Publisher: Mica Haley
Acquisitions Editor: Rafael Teixeira
Editorial Project Manager: Ana Claudia Garcia
Production Project Manager: Chris Wortley
Designer: Matthew Limbert

Typeset by Thomson Digital

Metabolomica e microbiomica. La medicina personalizzata dal feto all'adulto

This translation of Metabolomica e Microbiomica. La medicina personalizzata dal feto all'adulto (978-8-8986-3607-5), by Vassilios Fanos was undertaken by Elsevier Inc. and is published by arrangement with Hygeia Press di Corridori Marinella.

Translator: David C. Nilson

Contents

Author's biography

Vassilios Fanos is a full professor of Pediatrics and Director of the Post-Graduate School of Pediatrics of the University of Cagliari, the Director of the Neonatal Intensive Care Unit, Neonatal Pathology and Neonatal Section of the *Azienda Ospedaliera Universitaria* and the University of Cagliari. He is a pediatrician, neonatologist, and hygienist. He is the editor-in-chief of the *Journal of Pediatric and Neonatal Individualized Medicine* (www.jpnim.com), the official journal of the Union of European Neonatal and Perinatal Societies. He is a member of the International Perinatal Collegium and the president of the National Risk Management Committee of the Italian Pediatrics Society. He is a board member of the Italian Society of Neonatology and Chief of the Italian Society for Developmental Origins of Health and Disease. He is also a board member of the European Association of Perinatal Medicine and the chairman of its Educational Committee. He is an assignee of European and national research projects, the speaker and moderator at many international congresses and the organizer of congresses (among which the annual *International Workshop on Neonatology* in Cagliari, now in its 12th edition). He has organized and held refresher courses, has edited for numerous international agencies in the field of project financing, and refereed for more than 50 international medical journals. He has authored over 500 scientific articles, of which more than 320 are present in PubMed. Up to now, he has published 21 books as the author and editor for Humana Press (Springer), Mondadori, Franco Angeli, Agorà, Biomedia Source Books, and Hygeia Press. He is considered one of the major experts in the field of clinical metabolomics in perinatal and pediatric medicine. At present he is participating in more than 95 studies on metabolomics with researchers from 18 different countries.

Introduction

Research, supported by sophisticated technology, has taken giant steps forward in recent years and new languages have appeared in the medical field. The new technologies will enable us not so much to anticipate the future, but to make it possible. But what is important is not to lose sight of the quite peculiar significance of the meeting between physicians and patients and to continue to see things through the eyes of the patient.

In this volume you will find almost all, or at least much, of the new knowledge and languages of medicine: systems biology, systems medicine, network medicine, and medicine of complexity. What emerges ever more clearly from research is that we are individuals within an ecosystem inhabited by a huge number of bacteria that guide and orient us and which we must learn to know and respect.

We speak of nutrition on many pages in the different chapters. Many diseases arise, in the presence of a genetic predisposition, from the encounter of a certain kind of diet with a certain kind of intestinal microbiota, a sort of "additional organ" that we often overlook. It is composed of bacteria present in our intestine: only the encounter of diet and microbiota can form the metabolites that pass into the organism and control and govern all our organs, including the brain. The good news is that by changing our diet and/or modifying our intestinal microbiota (with prebiotics, probiotics, symbiotics and tomorrow with a fecal microbiota transplant) we can intervene (or try to) concretely.

Many diseases in adulthood can be prevented by means of the future mother's proper diet, one that places the fetus in the best conditions for development. Diabetes, obesity, and autism are examples of diseases that could be reduced with this approach. Moreover, a proper maternal diet also makes it possible to reduce the incidence of preterm deliveries.

We have an extraordinary interindividual variability that goes from fragility to resilience, to antifragility. More knowledge on this must be acquired, not only by health professionals, but also by patients, the mass media, and the entire society.

Technological advancements and their application do not diminish the role of physicians: on the contrary, they represent a formidable instrument for extending their diagnostic potential and making possible 10 P medicine: personalized, perspective, predictive, preventive, precise, participatory, patient-centric, psycho-cognitive, post-genomic, public. The final goal is to improve the wellbeing of each patient rather than limiting our efforts to treating the disease only.

These new acquisitions and frontiers may be upsetting, at least at the beginning, but on reading this book at your pace you may find them more natural than you thought at first glance. In any case, this is the road to the future. Metabolomics and microbiomics will strongly influence medicine in the immediate future.

I have been passionately involved with metabolomics for 8 years now, together with my co-workers; without their enthusiasm and dedication our studies would not have been performed. For this reason I have thanked them all at the beginning of this volume.

This book has been in my mind for about 30 years, it was written in 3 months and can be read in 3 days. Passion makes the ordinary extraordinary and is contagious: in this sense I hope you will be infected.

Medicine centered on the person

1

Through the patient's eyes

There is a great difference between medicine centered on the person and that focused on disease. In recent years we have witnessed radical changes in the world of medicine and, to sum up what has come about, we can say that observing things through the patient's eyes appears today to be the only possible way to meet the challenges of medicine. To be more specific, we have gone from paternalism to a therapeutic alliance, from curing patients to caring for them, from the concept of sickness connected to a disease of the body to that of illness which considers patients in the wholeness of their beings with bodies and minds, from compliance of patients who "tolerate" what the doctor proposes to adherence of patients who accept the doctor's proposal. This itinerary marks the passage from a physician-centered system to a patient-centered one. This sees patients as being less and less passive, more and more active and assuming a leading role in their improvement (empowerment). Furthermore, communication has assumed a growing importance in the relationship between doctors and patients: it is increasingly important to communicate "with" patients rather than simply communicating "to" them. Good communication is the key to arriving at patients' agreement with the physician's opinion. It is sometimes necessary to have a good relationship with patients' families as well. In the case of the pediatrician, contact with the patient is mediated by the parents, but also in the case of adult patients the family can provide valid support. In any case, in the relationship between physician and patient, an authoritarian attitude is becoming less and less acceptable. Readiness to explain to patients what is happening and the reasons behind therapeutic choices with a language that can be understood, especially by those with no medical training, leads to a shared plan for action. To establish a relationship of this kind with patients it is above all necessary to know how to listen.

Table 1.1 illustrates today's new tendencies which are changing the panorama of health care; however, other profound changes will take place in the next few years, when the significant improvements in basic biological research, associated with the impact of technological advances and the possibility of analyzing enormous amounts of data, will bring about revolutionary changes in medicine, with immediate and positive fallout for our health and a rapid increase in life expectancy.

An irresistible evolution in medicine

An irresistible evolution in medicine, which will go far beyond the limits of our imaginations, will be represented by the development and interrelation of four technologies: nanotechnologies, the area of technology "... concerned with manufacturing to

Metabolomics and Microbiomics. http://dx.doi.org/10.1016/B978-0-12-805305-8.00001-7

Table 1.1 Relationship between physicians and patients: a scenario being profoundly altered

Past physician–patient relationship	Future physician–patient relationship
Paternalism	Therapeutic alliance
Treatment	Care
Disease	Illness
Compliance	Adherence
Patient passive	Patient active (empowerment)
Physician-centered medicine	Medicine centered on patients and their families
Communicate to...	Communicate with...

Modified from Fanos (2008).

dimensions or tolerances in the range of 0.1 and 100 nm [that plays] a key role in many areas..." in Albert Franks' words; biotechnologies, information and communications technology, and cognitive sciences. Although we are not yet fully aware of it, the human race is about to enter a new stage in its evolution, a true revolution that begins with today's reality in an ongoing metamorphosis.

But we must be careful, and it is not by chance that I say this at the beginning of this book, not to lose sight of the quite special nature of the coming together of physicians and their patients. The risk of future medicine will be that of appearing as miraculous and infallible since it is based on scientific evidence supported by sophisticated technology; but if it is separated from patients and their needs, medicine will run the risk of losing sight of its limits, which will objectively continue to be present in consideration of the extreme complexity of the human being. It must be clear that there is no project, architecture, design, or product, whether material or intellectual, that can be compared to the complexity of human beings and their lives.

The five great ideas of biology and medicine

What is life?

What is life? A very simple question, an impossible answer. The Nobel laureate Erwin Schrödinger delivered a renowned series of lectures in 1943 entitled with these very words. It was followed by publication of a short book that sparked a lively scientific debate.

In recent years giant steps have been taken compared to the past and now we are immersed, as Craig Venter says, in the digital age of biology. But life is extraordinarily complex and we are still far from answering the question: what is life?

In the history of medicine an enormous number of theories, hypotheses, and opinions have been put forward. But if we were in a hot-air balloon with everything that has been written over the centuries and were in danger of falling, we would be forced to throw overboard most of the proposed answers. We would keep the important ones to the last, those which according to the Nobel laureate Paul Nurse are the five great ideas of biology and medicine: genome, cell, biochemistry, evolution, and systems biology (Table 2.1).

What are we talking about when we speak of the genome?

What are genetics and the genome? When we speak of genetics we refer to the study of single genes, whereas genomics deals with an individual's entire DNA: the genome is the legacy we have received from our parents and which we can hand down to our offspring. On June 26, 2000, Bill Clinton informed the world that the Human Genome Project (HGP) had completed the sequencing of the human genome. The announcement was made in the presence of many scientists, including the Nobel prizewinner James D. Watson who in 1953, together with Francis Crick and Maurice Wilkins, had discovered the double helix of DNA, our genetic heritage.

Our genome is comparable to a text composed of three billion letters, which would cover about one million pages: a gigantic "user's manual" continuously read and interpreted by the cell that contains it. DNA is a sort of double string or, to be more precise, a double helix composed of the sequence of four chemical units called nucleotide bases: adenine (A), thymine (T), guanine (G), and cytosine (C), the four-letter alphabet of our lives. The order of the letters is of the utmost importance, just as it is in our writing, in which the four letters can spell different words: "tags," "stag," "gats." It all depends on the sequence of the letters. The instructions given by the DNA are essential for living an entire life, from the vital spark to old age.

Metabolomics and Microbiomics. http://dx.doi.org/10.1016/B978-0-12-805305-8.00002-9

Table 2.1 The five great ideas of biology and medicine

• Genome
• Cell
• Biochemistry
• Evolution
• Systems biology (or systemic biology)

The genes are the single chapters of the genome. Some cells express certain genes, others express different ones, in a well-planned and defined space-time sequence. For example, in kidney cells only the genes dedicated to the kidney function are activated, whereas in the pancreas only those dedicated to that organ are dynamic. The genome represents an important part of our lives and it would be wrong and absurd not to acknowledge this. The genes, although they play an indispensable leading role, cannot act on their own and cannot explain everything. For researchers all over the world it was a hard job to arrive at the identification of the genome, "the code of life," but now that we know it (even though we understand but a small part of its meaning) we realize that it does not represent us completely. Success in defining the genome was for human beings a "mission impossible," almost like finding the Holy Grail. We are not our genes, or better still, we are not only our genes, as was thought some 15 years ago. Genes alone do not explain why the brain has become what it is. The brain is our most important organ, so much so that we could say we *are* our brain. In reality, the brain matures by means of a close relationship between genetics and the environment, since the genome is constantly confronted by, and related to, the environment, and this is how our brain's connections form, which is to say our connectome.

The "selfish genes," which gave the title to a famous and controversial essay by Richard Dawkins, the first edition of which was published in 1976, now belong to a level of knowledge that, if not eclipsed, at least requires updating. Dawkins' vision of evolution identified the genes, rather than the individual or the species, as the subject of natural selection. Genes are "selfish" because only the ones with the highest probability of continuing to be replicated appear to be transmitted. But today the monadic concept of the gene as an independent unit has been integrated by the genome and, as we know quite well, the genome is in constant equilibrium with the environment.

Broadly speaking, we are going from a "DNA-centric" vision to one that is decidedly more complex, which only narrowly can be called "protein-centric" and which focuses attention on the cell.

The cell, the city of life

The cell is the biological foundation of life and represents the essential and irreplaceable unit of every living organism. Each cell holds within it the genetic heritage of an individual. In defining a cell, I believe the words of Richard Feynman, Nobel laureate for physics in 1965, are to the point: "Cells are very small but very active; they produce different substances, they move, they get excited, they do wonderful things, and

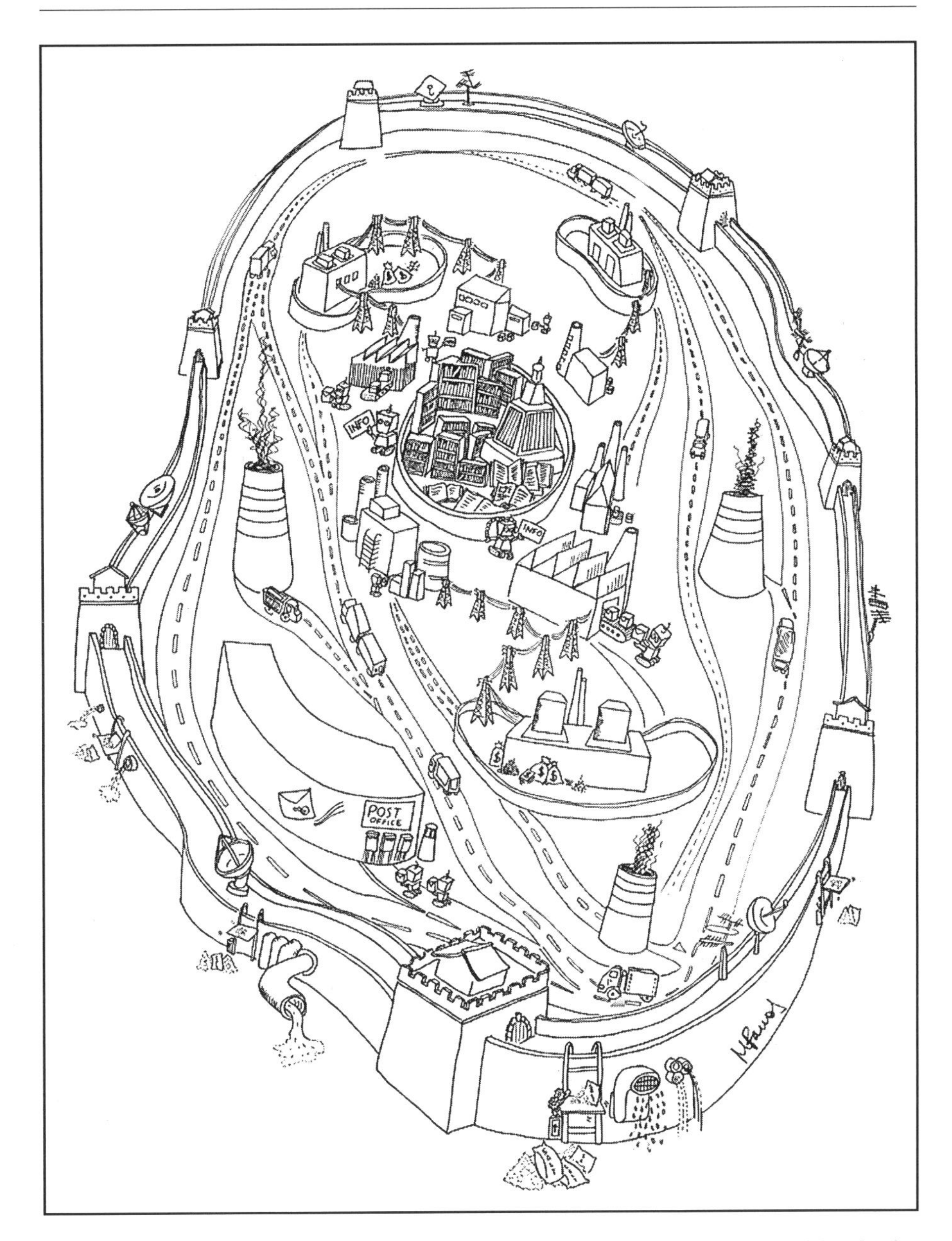

Figure 2.1 Schematic and figurative representation of a cell (Illustration by Margherita Fanos).

all on a very small scale." I find particularly apt the metaphor used by the physicist Peter M. Hoffmann in his book entitled *Life's Ratchet: How Molecular Machines Extract Order from Chaos*, to explain how a cell is structured, so much so that I took it, enlarged on it, and slightly modified it as follows (you will also find it translated into figurative language in Fig. 2.1).

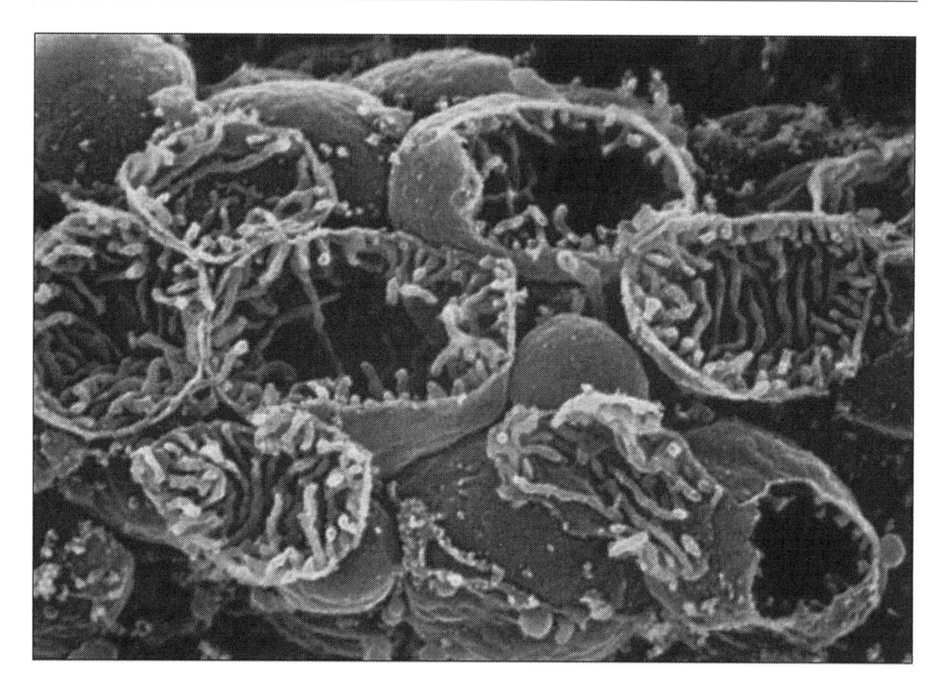

Figure 2.2 Interior of mitochondria (of the liver). We can see the mitochondrial crests (13,000×).
From Mocci et al. (2014).

A cell is like a city: it has walls (the cell membrane), a library that contains books (the nucleus that contains the genetic information), factories (the ribosomes), highly efficient power plants (the mitochondria, Fig. 2.2) which are cell recharging points; these "mitochondrial plants" use the energy provided by food to recharge the "energy currency" (the adenosine triphosphate, ATP), which is the money that pays the costs of energy, without which we would be unable to think, breathe, or move about. The cell has highways (the microtubules and actin filaments), trucks (the kinesins, which carry material within the cell), incinerators (the lysosomes), post offices (the Golgi apparatus, Fig. 2.3), and many other structures that perform vital functions, such as the pipe network and pumps (pores and ducts) that intake and output water and salt in the cell, or the antennas (the receptors) that communicate with neighboring cells and connect cells to the outside. When the incinerator breaks down, it is like pressing the cell self-destruct button: it goes beyond the point of no return and dies.

In Fig. 2.1, the walls surrounding the library represent the nuclear membrane, which in the cell surrounds the nucleus (Fig. 2.4). The nuclear membrane is equipped with pores, shown in Fig. 2.1 as doors from which emerge small robots that carry information for the production of proteins in the factories with the energy produced by the power plants and then transport them along the highways to where they are needed. Some of these robots are highly specialized and transport only one kind of cargo, for example, hemoglobin. The ribosome is the real choreographer of protein production.

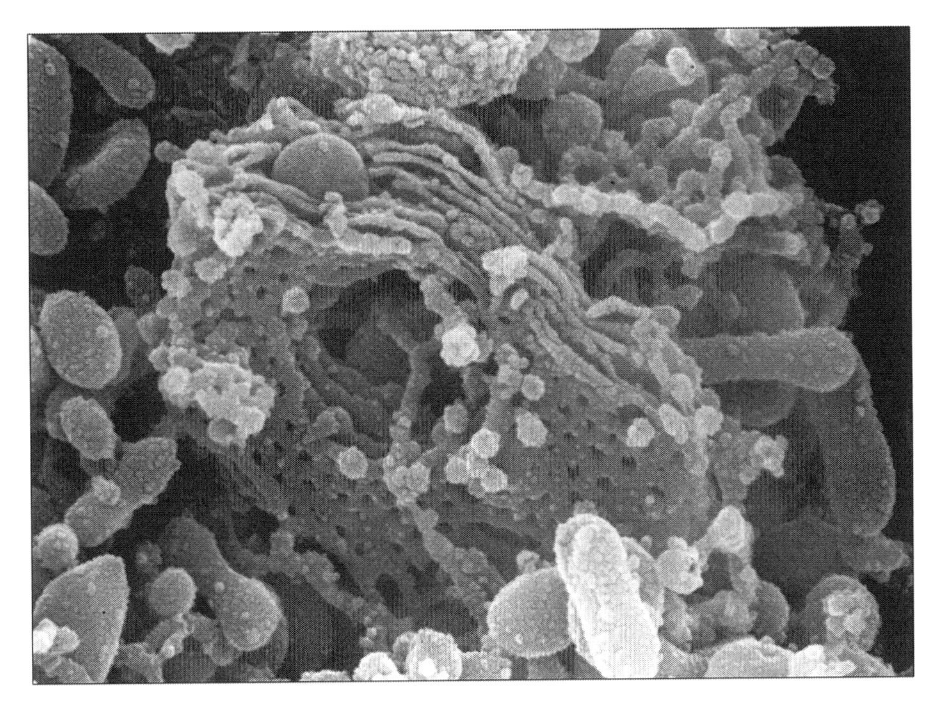

Figure 2.3 The Golgi apparatus (of the liver). It appears as a spongy body surrounded by vesicles containing proteins to be exported (15,000×).
From Mocci et al. (2014).

All these activities are performed by molecular machines that regularly consume ATP, which is created by ATP synthase, an extremely small enzyme (about 200,000 times smaller than a pinpoint) that is composed of 31 amino acids and performs about 60 rotations per second.

Ask yourself the following question, since it will be useful later on: Isn't it singular that our cities are substantially organized like the cells of our organism?

Biochemistry: maximum yield, minimum cost

Everything that occurs in all living organisms is functional to production, preservation, and increase of energy by means of biochemical processes. We can say that under physiological conditions the organism follows the mantra: *maximum yield, minimum cost*. This situation changes in the case of illness, as occurs in an inefficient engine subject to wear and tear in which heat is dispersed and fuel produces less mechanical energy. In many sicknesses the mitochondrial power plants are strongly involved and their involvement is evident in the determination of congenital malformations and adult diseases that begin, as we shall see later, already in the womb. Furthermore, mitochondrial dysfunctions and defects of the oxidative metabolism are characteristic

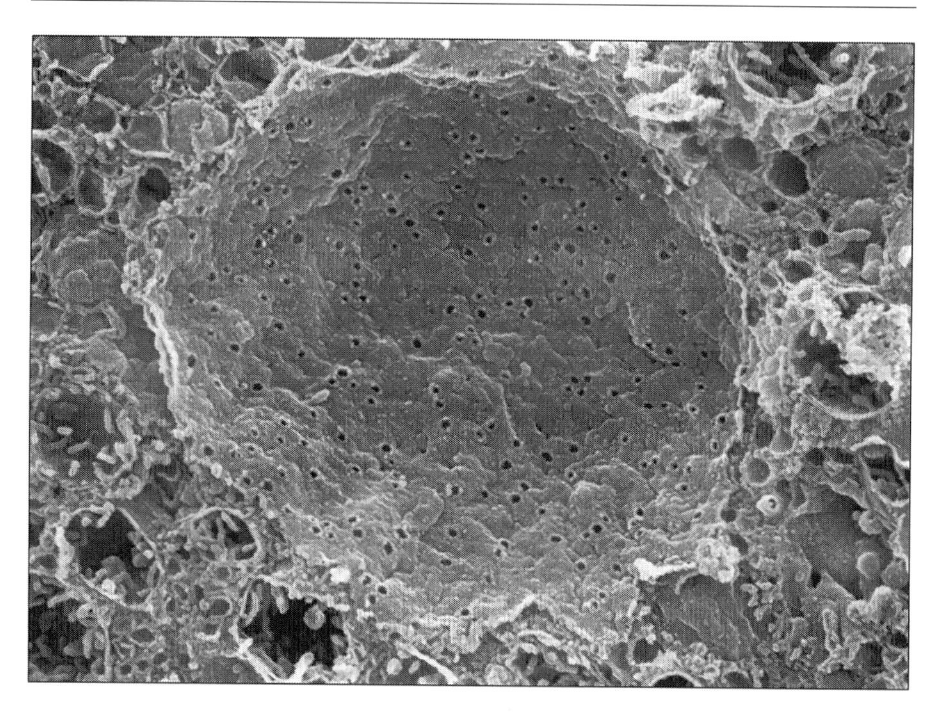

Figure 2.4 Interior of the nuclear membrane (of the liver). We can clearly see the nuclear pores (10,000×).
From Mocci et al. (2014).

of many chronic diseases not usually classified as mitochondrial diseases. This is true for neuroimmune pathologies and frequent, serious neuropsychiatric disorders such as autism, bipolar disorder, schizophrenia, major depression, multiple sclerosis, Parkinson's disease, and the chronic fatigue syndrome.

Evolution: I change, ergo I survive

Darwin was a timid, conservative man who asked upsetting questions. The evidence that Darwin's theory is not just a theory is overwhelming. Evolution has always been a fascinating concept but today, compared to the past, it is even more crucial for human wellbeing, medical science and in understanding the world around us. Health depends on the ability to resist physical, mental, and social stress: those who cannot resist stress or resist it inadequately are fragile, those who can stand up to it are resilient and, in Taleb's recent definition, those who resist stress and change are antifragile in the sense that they become fitter. In an article that appeared in the Italian daily *Corriere della Sera*, Taleb's philosophy was defined as "twenty-first-century Darwinism." Darwin was farsighted in teaching us that it is not the strongest, nor even the most intelligent species that survive, but those that are predisposed and ready for change.

Systems biology: the whole is not equal to the sum of its parts

The fifth great idea of biology is systems biology, the biology of systems or systemic biology (lately we speak more specifically of systems medicine), a theory perceived as "complicated." In fact, its key concept is the complexity of every living organism. Considering humans, this complexity reaches incredible levels.

From the practical standpoint, some complex problems in nature are unsolvable with the scientific instruments we have at hand today since, as the geneticist Edoardo Boncinelli has pointed out, complex systems have too high a number of variables and a discontinuous and unpredictable behavior.

Aristotle was aware that the whole is not the sum of its parts. Placed in our context, this means that an organism's behavior cannot be predicted through the properties of its single components. Understanding the single parts is crucial and strategic, but the parts are not always enough to gain an understanding of the whole. It would be like saying that it is important to disassemble an engine to understand what it is for and how each single piece works, but we cannot say that having all the pieces means that we can reassemble it and make it work. Complex interactions, although generated by simpler underlying parts, turn out to be independent in any case.

The study of the whole is what we call *holism*, whereas *reductionism* is the study of the single parts. In today's scientific panorama there is a widespread ongoing debate between upholders of the holistic approach and those who esteem the reductionist approach. In reality, like *structure* and *function*, *holism* and *reductionism* are bifrontal Januses, the two sides of the same coin.

In the holistic approach we find the so-called *-omics* disciplines, from the suffix *-oma*, which in a spreading convention denotes the totality of a word's root. These disciplines, such as genomics, transcriptomics, proteomics, and metabolomics, photograph the complexity of biological systems and are candidates gaining ground to replace traditional laboratory methods, which are less sensitive and less specific in diagnosing disease. The -omics are capable of identifying the single subject, be it normal or pathological, with the simultaneous analysis of enormous amounts of data, with samples often collected nonivasively. This analytical capacity is what is being increasingly defined as *data-driven intelligence*. To reach this requires an interdisciplinary approach: scientific knowledge is now so vast that researchers may overlook important connections, not because they are subtle or difficult to perceive, but because nobody has vast enough scientific knowledge to see such connections: in a gigantic haystack it may be hard to find a needle, even if it's ten yards long. An interdisciplinary approach in science is needed to solve the great problems of human beings.

The medicine of the future

Medicine of the future, beyond all imagination

The complexity of biological systems is emerging forcefully in the most recent literature. Table 3.1 presents what I personally believe to be the state of medicine today and what it will be in the future.

From epidemiologic to individualized medicine

Today's medicine is epidemiological. Patients go to their doctors and hear, for example, that they have a 25% (or one out of four, if the doctor is a good communicator) probability of having, or not having, a problem or a disease. However, patients respond that they are not interested in statistics, numbers, and percentages: they want to know what will actually happen to them. Mothers and fathers will tell the doctor that they are interested in Frank and Laura, their children, and not the children of others.

A medicine that treats patients with a given disease as being all the same but does not know the best treatment for a given individual is of little use to patients: disease is an exquisitely individual condition and as such must be approached. The treatment of single individuals calls for individualization and personalization of treatments.

From descriptive to predictive medicine

Today's medicine is descriptive. Hippocrates had observed all this more than 2000 years ago. For example, as concerns the kidney, he had substantially described infective, inflammatory, tumoral and stone-related pathologies. Future medicine will succeed in predicting what will happen to Frank and Laura, not only in describing what has already taken place. It is predicted that in just a few years health care systems will reimburse hospitals only for what they spend on medicines that are truly effective and without side effects. It will therefore be necessary to acquire the most up-to-date technologies such as pharmacogenomics and, at a later date, pharmacometabolomics, which I will write about later on.

From reductionist to holistic medicine

In health care today, we follow a reductionist approach. We are capable of performing basic laboratory tests which, although they seem to be many, in reality are few, but medicine is moving in the direction of a holistic science. We will arrive at the simultaneous dosage of hundreds, if not thousands, of biomarkers in different organic fluids. The number of laboratory tests we perform today in our hospital laboratories is

Metabolomics and Microbiomics. http://dx.doi.org/10.1016/B978-0-12-805305-8.00003-0

Table 3.1 Differences between today's medicine and that of the future

Today's medicine	Future medicine
Epidemiological	Individualized
Descriptive	Predictive
Reductionistic	Holistic
Reactive	Perspective
Based on genetics	Based on epigenetics

Modified from Fanos (2012).

a drop in the ocean: fewer than 100 markers are dosed compared to the 150,000 that have been studied up to now. An ocean of biomarkers in a drop of urine or blood will be what we will be able to monitor thanks to the holistic approach: we will have an all-in-one tool, a diagnostic instrument that will include everything.

From reactive to perspective medicine

Today's medicine is reactive, that of the future will be perspective. At present, laboratory tests reveal disease when it is already in progress and, as alarm systems, they leave much to be desired. The diagnosis of a tumor today can be pronounced only with some delay, after it has already developed. Another example: if in the blood we find an increase in creatinine, a marker of kidney damage, this tells us that there is an ongoing renal injury. But when the increase has taken place and I can measure it, it is already too late; I have lost an important amount of nephrons, the "workers" that toil untiringly around the clock to ensure the kidney's function. Are we interested in continuing this state of affairs as patients, patients' parents, and as health providers? What will interest us more and more in the future is an answer to the following questions: what do we mean when we speak of the state of health and wellbeing? How can I know, improve, increase, and defend my wellbeing against the onset of a tumor or any other disease?

From medicine based on genetics to that based on epigenetics

DNA is a code that represents the fundamental characteristic of every living organism. With the extraordinary amount of information contained in our genome, molecules know how to organize and where to go. But the genome is not enough to predict everything and plan everything in detail, as is normally thought. If most of our characteristics are determined by genetics, how is it that people with the same genome can appear quite dissimilar? The answer lies in epigenetics, a term that comes from the word *genetics* with the Greek prefix *epi-* (over). If the genome is like a large cookbook, the epigenetic mechanisms are the bookmarks that show which pages to go to. They are complex mechanisms that intervene with biochemical processes of methylation (keep in mind the initial *m* and think of mortification, extinction) and those of acetylation (keep in mind here the *a* of activation). They act on the genome as a sort of control room. Epigenetic mechanisms are reversible and can change over time. A quite effective image is that of a system of "switches" that activate and deactivate the genes in response to environmental stimuli such as stress or nutrition. Each gene remains the same, but it can go through different epigenetic states (switched on, switched off) and at different times.

This state of affairs introduces a further element of great variability and complexity and leads to potentially enormous individual differences. The proof of what I am saying is represented by identical twins, of whom I will speak later on: they have the same identical DNA, but different epigenetic profiles. The incomplete sameness of identical twins in terms of health indicates that the genome cannot predict individual results. It is important, first for all physicians and health care providers and then for patients, to understand that we have evolved as a human species to such a point that there is an extraordinary interindividual variability between one person and another. I will return to this on the following pages.

Thus the role of epigenetics emerges strongly. A quip is circulating in the academic world: if at a scientific congress on medicine somebody asks you a question you cannot answer, just say with no hesitation, "It's epigenetics." However, epigenetics is certainly not the fashionable answer to everything we do not know, and the idea of a self-sufficient epigenetics is wrong, just as the emphasis placed on DNA was wrong: genetics and epigenetics are closely interlaced. Structurally, the genome is comparable to the hardware of a computer and in practice it determines the confines of what it is possible to do, whereas the epigenome acts as the operating system without which the genome does not receive the instructions it needs to operate.

While lecturing on this subject I saw that examples can help in understanding the interactive mechanisms between genetics and epigenetics. Here are some of these.

- Genetics is written with a pen and cannot be cancelled, whereas epigenetics is in pencil and can be changed.
- Genetics proposes, epigenetics disposes.
- Genetics is a lengthy piano keyboard with about 25,000 keys (our genes). Epigenetics decides the choice and succession of the keys: which keys play and in what sequence. What can come out is Mozart's Symphony no. 40 or Stravinsky's Firebird, with a quite different impact on listeners.

We cannot change our past (the genes inherited from our parents) but we can try to direct or change our future by modifying the epigenetic factors, for example, by following a healthy diet, protecting ourselves from environmental pollution, avoiding, limiting, and personalizing the use of medicines and modulating our lifestyles.

As concerns epigenetic factors, the most important of these is nutrition. Preventive strategies can be adopted by changing our diets and lifestyles. Thomas Edison, the inventor of the light bulb and more than a thousand other patents (to give you an idea, this means one patent a week for 20 years) predicted many years ago that future doctors would cure and prevent diseases through diet.

In the fields of pediatrics and neonatology, diet is even more important because the child is a dynamic organism undergoing constant growth. It is enough to observe the development in height and weight in the first year, or the growth and morphological and functional maturation of the encephalon in the first 3 years. Such important changes in so short a time take place at no other age. Moreover, a mass of experimental, clinical, and epidemiological data of increasing importance highlights the key role played by perinatal programming in determining the good or poor health of an adult. Everything, or almost everything, is decided in the first periods of life or even in the womb. The susceptibility to many diseases begins in this very period of life and their prevention begins during pregnancy.

DNA is not everything: not even identical twins are identical

I repeat: identical twins are not identical. The comparison of identical twins is enlightening because they represent the "perfect experiment in nature": they have the same DNA and an appearance that makes it hard to tell one from the other. I knew two male twins I will call M and G for simplicity's sake. When young, they were like two drops of water and even exchanged girlfriends. If M's girlfriend said, "Take me where we went last night," but she was then with G, he would say, "Why not go someplace else?" if the brothers had forgotten to exchange information. This was before cell phones were invented.

Let's go back a step. There are two kinds of twins: monozygotic (also called identical twins, born from a single fertilized egg, whose genes are all the same) and dizygotic (born from two different eggs and in this case are fraternal). Here I want to deal with monozygotic twins, who should be identical but are not. Here are some examples.

C, a boy of 14, does well at school, he wears a jacket and tie, he keeps his hair short, has been going with the same girl for almost a year, he will marry her and they will have four children and live happily ever after. His identical twin, S, dresses shabbily, has a poor record at school, is antisocial and violent.

F, an adult of 48, is a brilliant professor in the medical school of a prestigious university, he is invited to lecture all over the world and has patented several inventions. His identical twin, R, has been arrested many times for drunken driving and considers himself a failure; his wife has left him and his children are ashamed of him.

L, a woman of 62, has Alzheimer's disease with an early onset whereas B, her identical twin, shows no sign of this disease and takes care of her sister.

How can it be that these identical twins are by no means identical? Identical twins, born with the same DNA and, even though raised in the same family environment, can become surprisingly different with the passing of the years. Older twins show great differences in the overall content and distribution of their epigenetic characteristics, with patterns of gene expression that are unique. This is epigenetics! The differences are more important than the similarities: they derive from the interaction between DNA and environment, and are only partially inheritable.

We must therefore conclude that environmental experiences, sometimes even casual, modify the expression of our DNA by introducing new tags in the genome for new transcriptions: they do not change the DNA, but can activate or silence (or if you prefer "shut up") genes to different extents. Our actions become new transcriptions: think, for example, of nutrition or maternal stress during pregnancy, or stress in the child's or adult's life.

The major epigenetic differences have been observed in twins who have spent less time together or have a different medical history. Environmental factors (smoking, pollution, diet, and physical activity) can impact on epigenetic profiles and may anticipate or delay the onset of diseases in identical twins. Our genetic legacy can produce relatively different results depending on the kind of epigenetic regulation that sets in as a result of environmental and internal stimuli. Eating in a certain way, exercising or not, being or not being loved when in the womb and as babies, living in a more or less polluted environment, having or not having a chronic disease—in short, real life—results in an epigenetic regulation of our genome.

We can say, more properly and briefly, that identical twins have the same genome, but a different epigenome, a different phenome (one has a disease whereas the other does not), a different diseasome or state of health (one has a moderate form and the other a serious form of the same disease).

Epigenetic mutations take place already in the uterus and accumulate throughout life: thus the differences between the two individuals, who at conception were identical, increase as they age. Let's take another step forward: already at birth identical twins may differ greatly in weight as the result of events in the womb that cause an asymmetric supply of blood, nutrients, and oxygen. This is intrauterine epigenetics.

Particularly interesting is the analysis of Fig. 3.1: identical twins share certain pathologies to underline the strong impact of the invisible hand of heredity. If a person

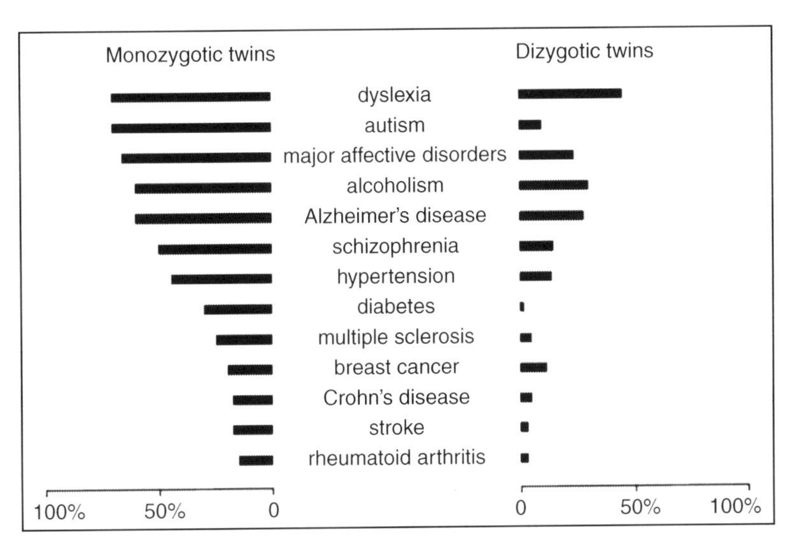

Figure 3.1 Mean probability of sharing certain pathologies measured in identical (monozygotic) twins and fraternal (dizygotic) twins that suggests the strong influence of heredity.
Modified from Miller (2012).

has a disorder of the autistic kind, the monozygotic twin has approximately a 70% probability of manifesting the same disorder compared to just 5% for a dizygotic twin. But here we are dealing precisely with probability, which epigenetic factors can change, even disrupt and subvert. At present I am participating in a European 7th Framework Programme for Research and Technological Development (ACTION—Aggression in Children: unraveling gene-environment interplay to inform Treatment and InterventiON strategies), with twelve partners from eight different countries coordinated by Dorret Boomsma, a professor of biological psychology at the VU University of Amsterdam and founder of the Netherlands Twin Register in 1987. By means of the -omics disciplines the project studies social aggressiveness in identical twins who display quite different behavior.

We can conclude with the title of Tim Spector's excellent book on twins *Identically Different*. He is a professor of genetic epidemiology at King's College of London who describes himself as one of the many scientists who took for granted the genocentric vision of the universe (genes at the center of all) only to discover, through an intense and prolonged study of twins, what our genes do not control: we are so complex as persons that our diversity cannot be explained only by our genes. The number of genes we thought we had (a hundred thousand or slightly more) has been reduced to about 25,000, more or less the same number possessed by a lowly worm.

On the new farm... there's a sheep, a cat, mice

The importance of studying and accounting for the interaction of genes with the environment is also demonstrated by cloned animals.

The well-known sheep Dolly was the first mammal successfully cloned by back-programming an adult cell until it became a universal cell. This was a great success that resulted from enormous efforts: a bullseye after 277 shots, incredible costs, premature aging and death of clones obtained, ethical problems the size of Everest. Dolly's DNA was exactly the same as her mother's.

Less famous in the "new farmyard" was the cat Rainbow and its cloned copy called CC (aka Copy Cat). Although genetically identical, the two cats do not appear identical nor do they behave the same way. The most evident disparity is the color of the fur, gray–orange in Rainbow and all gray in CC. The differences in behavior are even more pronounced and the personality is where we see striking differences, since a myriad of factors intervene and interfere. The influences, positive or negative, start in the womb and the mother's diet is decisive: it has been shown that a mother's undernourishment is associated with evident brain defects in the child and with abnormal levels of aggressiveness in the long run. Even when the expectant mother is the same, the intrauterine environment may undergo changes. A female kitten closed between two males in the womb is immersed in a "testosterone bath" which leads to a masculinization of its brain. Thus a clone is never truly identical to another because many environmental factors come into play. If you have cryopreserved cells, for just six thousand dollars you can clone your dog Ringo or your cat Fluffy, but the clones will differ significantly from the original you kept in your house or garden.

All this was predicted in the film *Multiplicity*, distributed in 1996. Doug, the hero, stressed by his job and family, accepted to be cloned. At the beginning, everything seemed to proceed for the best: the clone worked for him, which let him spend his time as he liked, but he soon realized that one duplicate was not enough and he was cloned a second time. The first clone, he too distressed by his job, decided to duplicate himself and produced the clone of a clone. But all these clones were not perfectly identical one to the other and not even to the original: each copy was different and had his own psychological imprint. These differences led to further problems instead of solving the original ones: the hero could no longer stand his clones, even became jealous of them and sent them away from him and his life.

The last example is a classic of epigenetics: the Agouti mice. These mice have a mutation (A^{vy}, Agouti viable yellow) involving the Agouti gene, on which depends coat color: yellow mice instead of ones with the normal dark color are born. They are always hungry and are destined to become obese and diabetic. But not all the mice with the same mutation are yellow or obese: some are darker in color and thinner. We know that the introduction of supplements rich in methyl groups (such as folic acid and vitamin B12) into the diet of yellow Agouti mice during pregnancy influences the coat color of the offspring, depending on how their Agouti gene is methylated during embryonic growth. Moreover, the Agouti mice represent one (up to now) of the few proofs of how phenotype variability induced by epigenetic variability can come to the fore in later generations.

The "-omics" technologies

4

The new languages of medicine

Once more we go to Aristotle, according to whom the whole is not represented by the simple sum of its parts. Applied to our context, this means that the behavior of complex systems cannot be predicted from the properties of their single elements, exactly as we saw previously for cells.

We have become used to the idea that a medicine based on scientific evidence, such as the numbers that emerge from the laboratory, is an exact science like mathematics, without realizing that mathematics and statistics can provide us only with a representative model of reality, which is not reality itself. The isolated result of a laboratory test does not entirely represent the reality of a living organism, which is far more complex. To arrive at a better and more reliable picture we need an enormous amount of interconnected data and the science that studies living beings is now oriented in the direction of systems biology, which integrates multiple levels of information to arrive at an overall knowledge of biological systems.

So what are these new languages of medicine? I have listed these in Table 4.1. If readers, especially health providers, are not acquainted with at least five of these terms it means that they need more information and training. During a meeting, at the end of my lecture during which I presented this table as a slide, the director of a laboratory who was moderating the session commented: "I understand one thing: if I don't bring myself up to date on these subjects, I'm going to lose my job in 4 years, when my contract expires!"

-Omics technologies

Genomics, transcriptomics, proteomics, metabolomics (but about a hundred more subdisciplines, such as interactomics and fluxomics) are the so-called -omics disciplines. They use specific scientific acquisitions to comprehend an overall biological system from a holistic standpoint (systems biology, systems medicine).

In the field of systems biology the -omics disciplines propose a holistic approach to gain an understanding of the several molecular components (genes, transcripts, proteins, and metabolites) of a cell, a tissue, or an organism that play a role of increasing importance in clinical practice and the providing of health care.

Metabolomics and Microbiomics. http://dx.doi.org/10.1016/B978-0-12-805305-8.00004-2

Table 4.1 The new languages of medicine

• Systems biology
• Systems medicine
• -Omics technologies
• Network medicine
• Data-driven medicine
• Big data
• Fragility, resilience, antifragility
• Holistic medicine
• Individualized medicine

-Omics technologies: genomics

The first in time to develop was genomics: in 2010 the great Genome Project reached the age of 10 years and, in the meantime, its limits had also emerged. In the Italian daily *Sole 24 ore*, an article that appeared in 2010 under the title *Mappe del dopo genoma* (Map of after the genome), stated that research on the genome could be compared to the "charts in possession of the first navigators: sufficient for sailing round the globe but far from being a trustworthy representation of the world" and from today's navigation instruments. Despite the massive investments in research, the practical consequences of the project on health were modest and far below expectations in consideration of the funds invested. Many of these expectations were connected with the study of monogenic diseases such as Huntington's disease, cystic fibrosis, and sickle-cell disease, caused by mutation of a single gene. But these diseases represent only 2% of all genetic disorders. Most diseases involve more than one gene and factor, with an origin that is far more complex and articulated. The 30 genes up to now associated with obesity explain only 2% of the disease. Genes are not omnipotent: our complexity and diversity cannot be attributed to genes alone.

Some years ago I was a moderator at an international convention on genetic psychiatry. What surprised me most was that the different speakers, on brilliantly presenting their studies, classified a series of problems at the cerebral level on the basis of the number of genes involved. During the discussion, I wanted to know the role of environmental factors, such as mother's health, asphyxia at birth and, most of all, weight at birth. All those present were rather surprised: not one of them had considered these issues. It seemed that environmental factors did not exist and that everything was written in the great book of destiny, like in the film *Gattaca* by Andrew Niccol. Can metabolomics tell us something more?

-Omics technologies: metabolomics

Metabolomics, also called the "new biochemistry," is an approach based on the systematic study of the complete set of metabolites in a biological sample. Other definitions, such as "metabolomic profile," "metabolic fingerprint," and "metabolite target analysis," are commonly used in current practice. The term "metabolomics" comes from the Greek μεταβολή (the same root as "metabolism"), which means "change," and the suffix *-oma*, seen previously, which means "all." It is the global study of metabolites, small molecules present in cells, tissues, organs, and biological liquids. They are lipids, glucides, small peptides, amino acids, and vitamins. Taken all together, metabolites make up a metabolic fingerprint, a sort of metabolic sign specific to every individual. If I collected a urine sample from each of my readers, metabolomics would be capable of distinguishing one from the other with no superimpositions or errors. It is somewhat like discovering a metabolic "face," one that we could attach to our electronic health identity cards. For no other animal is the face as distinctive as it is for humans. Our extraordinary biological variability makes it possible to distinguish one person from another and this same variability is responsible for our many metabolic and metabolomic differences.

We could ask: are we going from the "egoistic genes," which are immutable and dedicated to self-preservation, to the "altruistic metabolites," which can change depending on environmental influences and are generous in offering suggestions that help us in orienting our choices concerning health?

The metabolites present in our bodies or eliminated from it, especially those excreted in the urine, assume a new role in decoding the complex future of our organisms: elements once considered marginal have now become the fundamentals of the new technology which, on close examination, enables us to delve into and expand an idea that is not new in medicine. The urine wheel, presented in 1506 by Ulrich Pinder in his book *Epiphanie Medicorum*, is a sort of proto-metabolomics: the combination of color, odor, and taste of urine was used in diagnosing disease. And here I wish to emphasize a further strength of metabolomics: this technique, with such a strong potential for providing information, can be used on samples collected noninvasively, such as urine, saliva, and hairs.

Metabolomics: the rebirth of science or Harry Potter magic?

Metabolomics, owing to the difficulty and cultural innovation that this approach requires, is often the target of suspicion and criticism. It is certainly not easy to go from an epidemiological conception of medicine to its personalization or from a descriptive to a predictive medicine: it is a question of changing our perspective comparable to the change brought about by the Copernican revolution and certainly not by magic tricks.

Already in 2005 the Massachusetts Institute of Technology of Boston had identified metabolomics as one of the 10 emerging technologies, stating that it would contribute to the earlier, faster, and more accurate diagnosis of diseases.

Two facts stand out: up to 2012, the index of PubMed (the most important medical database) contained about 10,000 works on metabolomics, with an exponential increase over the years; it is predicted that the global market created by metabolomics will expand at a compound annual growth rate (CAGR) of 26% from 2014 to 2020 and reach a value of $2.8 billion at the end of 2020. These data show the lively interest in metabolomics not only in the scientific world, but also among entrepreneurs, a demonstration that metabolomics is a discipline with solid scientific bases and a wide range of practical applications.

From the very beginning even the nonspecialized press has attributed to this new discipline an important role in the field of medicine for prevention, diagnosis, and treatment. Following are some examples I believe to be particularly important.

On February 3, 2008, the Italian newspaper *Corriere della Sera* published an article titled *La firma del nostro organismo* (Our organism's signature), in which metabolomics is defined as "a field of study in rapid expansion" and that: "It will be possible to trace a person's individual, unique and unrepeatable metabolic profile, one that promises to be extremely useful in medicine. No two metabolomes are the same, each one is a kind of very fine fingerprint."

In the May–June 2011 issue of the Harvard Magazine, in the article *Fathoming Metabolism*, the importance of metabolites in predicting future health at the individual level and providing answers beyond genomics is emphasized.

On October 31, 2012, in *La Stampa*, another Italian daily, appeared the article: *Scopri chi sei con le molecole. Inizia l'era della metabolomica: che cosa cambia, dalla diagnosi ai farmaci* (Discover who you are with molecules. The era of metabolomics has begun: what changes, from diagnosis to medicines). The article goes on to say that in a few years' time a third test may become decisive in the diagnosis of a disease. After the study of the genome and proteome, the new era of molecular medicine is focusing on a new discipline: that of metabolomics.

At the center of metabolomics is the important concept that an individual's metabolic state is the representation of that person's state of health or illness. Metabolomics is the foundation of personalized or tailored medicine, the only medicine possible in the coming years. The technology consists of two sequential passages: first, an analytical technique (usually nuclear magnetic resonance, gas chromatography-mass spectrometry, liquid chromatography-mass spectrometry) capable of determining a large number of metabolites present in biological samples; and second, multivariate statistical processing of the data obtained from the samples analyzed.

A patient's genotype defines the risk of contracting a certain disease or the probability of the reaction to a drug or the external environment, and in a certain sense it can be defined as a static profile. On the contrary, the phenotype reflects in real time the clinical reality more accurately at a given moment and can be defined in a dynamic profile.

The metabolome is so close to the phenotype as to be considered the phenotype itself: it is considered the most predictive phenotype and is able to take epigenetic differences into account. Genomics is "to be able to," transcriptomics is "to start," proteomics is "to do," and metabolomics is "to be."

Studying genomics is like knowing the catalog of 25,000 volumes in a library, comparable to our approximately the same number of genes. Studying metabolomics means identifying a list of 3000 volumes that are effectively consulted and knowing the profiles of those who consult them, perhaps even with a comment, a note, or a phrase on the features of the book, thus obtaining a far more dynamic and functional overview of the library. All genes are certainly important, but the most important ones are those we use and which interact with the environment, compared to those we do not use or use rarely.

A metabolomic investigation is like rummaging through a person's trash: we can understand almost everything, or at least quite a bit, about that person. If we find newspapers and magazines in the trash outside a private home we can get an idea of the leanings and interests of those who live there. If we often find cans of tuna fish we can imagine that the person is single and does not like to cook or does not have the time to do so. Depending on the brand of tuna, we can understand his or her economic situation, and even more by examining the wine bottle labels. If we find three prophylactics per week we know something about the person's sex life. And so on.

If you do not like the example of trash, we can consider the photographs in Mark Menjivar's *Refrigerators* project, which appeared in the *National Geographic* in December 2014: photos taken all over the United States of refrigerators with the doors standing open. Show me your refrigerator and I'll tell you who you are: we see the almost empty refrigerator of one who earns just over $400 a month, the well-stocked one of the obstetrician who buys only fresh local products, or that of the bartender who fills it with leftovers from his bar and does not even have time to eat them.

We are quite different at the beginning and we eat different foods in varying amounts. In the light of this diversity, let me anticipate a question I will discuss later on: how can we apply the same clinical protocols to all patients?

Extreme interindividual diversity calls for instruments that enable us to uncover this diversity, and metabolomics is an excellent response to this need. The metabolomic profile is a unique characteristic of all of us, one that is capable of identifying us with absolute specificity. It is like the photo on our identity cards.

Our metabolome also depends on its context and can vary with our physiological state and the state of development or pathology of cells, tissues, organs or the entire organism. It is like saying that we have a photo in which we are suntanned or have longer hair or pimples.

Furthermore, metabolic variations take place quickly: while some time may be required for a change in the expression of the messenger RNA, the response of metabolites can be seen in seconds or minutes, a circumstance that cannot be overlooked and that makes it best to focus our attention on the study of metabolomics rather than on the other -omics.

Here is an example to demonstrate the potentiality of metabolomics in turning research into a real clinical advantage for patients. A wide-ranging study was performed on the serum of more than 1300 patients in the search for a prediabetic condition by applying all four of the important -omics technologies: genomics, transcriptomics, proteomics, and metabolomics. What was the result? Seven genes, five proteins, four enzymes, and three metabolites were found, all connected to the prediabetic state. This result enables us to conclude that although we do not know these genes, proteins, and enzymes, the three metabolites alone are sufficient for a diagnosis. So we need not study all the -omics disciplines. It is enough to perform a metabolomic investigation only, thus avoiding organizational expenses and saving resources. A second practical corollary comes from the circumstance illustrated previously: the metabolome is the best marker of an organism's phenotype: if we find that there is a variation in the concentration of the three metabolites connected with prediabetes, we will have not only a probability or risk that the patient will develop diabetes, but also in all likelihood the certainty that it will develop. Knowing this, it will be possible to avoid, or at least postpone, the onset of the disease and its complications by immediately changing the epigenetic factors involved: nutrition, lifestyle, and so on.

This will certainty avoid in the future having so-called "risk patients" who need surgery or preventive therapies because they run the risk of developing a certain disease. While traditional laboratory techniques provide a probability of risk, with metabolomics there will be the certainty of risk or its absence.

Dogs, cats, and "electronic noses"

Some experiments, carried out also in Italy, have shown that dogs, if properly trained, are able to diagnose malignant prostate tumors with an accuracy of 97% by smelling a patient's urine (in Italy alone this cancer affects 25,000 men every year). Dogs are able to sense the special smell of urine of patients with prostate cancer that no human nose could ever detect. This may also be true for other tumors. What is the reason for this? Dogs possess more than 200 million olfactory cells compared to the 5 million of humans, which means that their olfactory capacity is much stronger than ours. Tumor tissues present a peculiar metabolism that produces hydrocarbons and high concentrations of nitrogenous compounds. They have a special smell that emerges at an early stage in the breath and urine of patients.

An even more striking experiment, published in 2007 in the leading *New England Journal of Medicine*, is that of the cat Oscar, capable of sensing the imminent death of the guests of a nursing home in Rhode Island: its sense of smell was probably able to perceive the presence of metabolites such as cadaverine, putrescine, sarcosine, or creatine even in quite low concentrations not ordinarily perceivable by the human nose.

The sense of smell of these animals works in a way similar to that of electronic noses, sophisticated devices capable of exactly revealing the percentage of substances contained in a sample on the basis of the odors they emit. Since odors are produced by the metabolites that develop in a substance, the working of electronic noses is based on the application of metabolomic analyses.

Metabolomic Pindaric flights: the most important applications of metabolomics

The most important applications of metabolomics are shown in Table 4.2. The metabolomic analysis of different biological liquids or tissues has been used with success in the fields of physiology, diagnostics, functional genomics, pharmacology, toxicology, and nutrition. Metabolomics photographs the route, the itinerary, and metabolic trajectory of a single individual, from birth to adulthood, from the state of health to that of illness.

The application of metabolomics in the field of metabolic screening at birth exemplifies this: since a temporal evolutionary trajectory of the metabolome from the first

Table 4.2 The most important applications of metabolomics in medicine

• Photograph of the physiological state
• Disease diagnosis
• Discovery of biomarkers
• Pharmacometabolomics: assessment of response to drugs
• Nutrimetabolomics: assessment of the effects of nutrition
• Phenotype classification
• Identification of metabolic pathways altered by diseases or therapies
• Functional genomics
• Characterization of natural products

day to the first week of life has been determined, if a neonate at birth has the typical metabolome of a baby of 7 days, or vice versa, this means that the child has problems, even when it appears to be in good overall conditions and has had no maladjustment problems at birth.

Just as pharmacogenomics and nutrigenomics are among the most important applications of genomics, pharmacometabolomics and nutrimetabolomics are now ready or near completion and will enable personalization of therapies and nutrition, respectively.

Evolution that becomes revolution: metabolomics as the generator of hypotheses

Metabolomics represents a revolution, or perhaps we can say that evolution becomes revolution. This technology is so revolutionary because it is the producer of hypotheses, and this is one of the reasons why metabolomics normally causes much perplexity among traditional clinicians. Some of them have compared metabolomics to fishing. An article in the literature responded with the following title: *Sometimes the fisher catches a big fish.* But it is traditional biochemistry that catches one fish at a time. Metabolomics is instead like catching hundreds of fish (metabolites) in a net and, by means of highly sophisticated methods, can say: "There's a kind of fish (or a small school of fish) so special and unique that I can understand from this just where in the world I'm fishing" (Fig. 4.1).

Four years ago in Istanbul, when the moderator asked at the end of one of my lectures: "Are there any questions or comments?", a colleague raised his hand and said: "I don't believe in metabolomics." I replied that I thought he was in the wrong room since we were speaking of science and not faith. A classic question among the FAQs that I receive when I present this subject concerns ethical issues, but I will speak of this further on.

Why all this perplexity and prejudice? I will try to explain more clearly. We are accustomed to the scientific method: we form a hypothesis and then prepare an experiment to see if it is correct or not. In the case of metabolomics, we do not need a starting hypothesis, and for this reason it can be used without a predefined target (untargeted

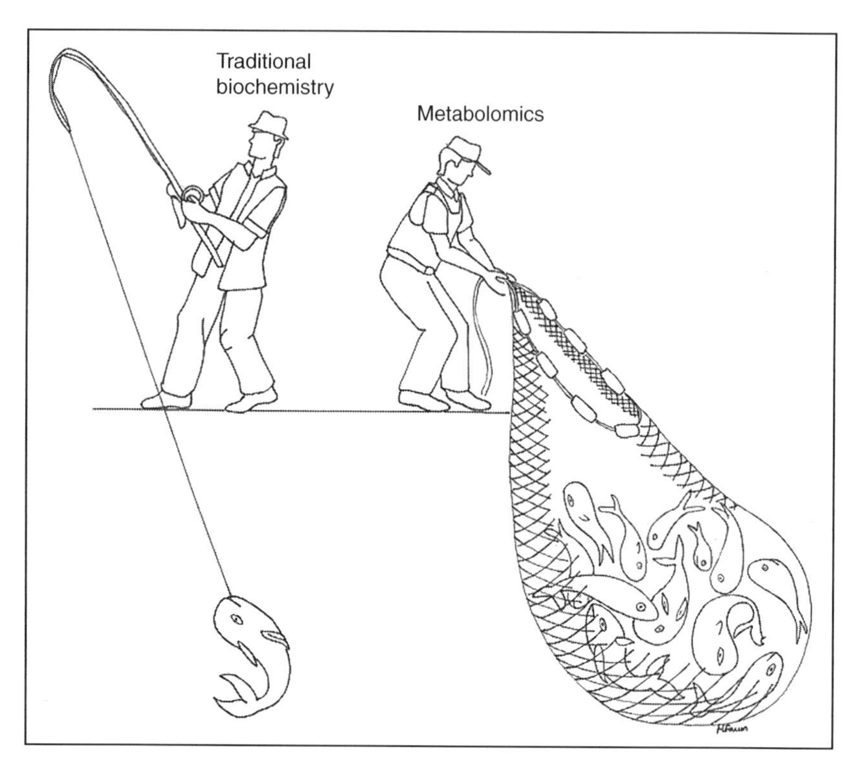

Figure 4.1 Exemplification of the difference between conventional biochemistry (left) and metabolomics (right). Traditional biochemistry is like catching one fish at a time, whereas metabolomics is like catching hundreds of fish (metabolites) in a net (Illustration by Margherita Fanos).

metabolomics). If we do formulate a starting hypothesis, that is fine, even better, but having it is not strictly necessary. Once we have determined and dosed hundreds of metabolites, sophisticated software places these metabolites in a descending hierarchical order, starting from the ones that best discriminate an individual before and after a change in diet or drug administration, or distinguish a healthy person from one who is ill. Taking the metabolites at the top of the hierarchy, I have a packet of discriminating metabolites. The advantage of considering all together the most important metabolites in descending order is extraordinary: it is like finding an algorithm with which to win the football pools. It gives us a unique vision of every process investigated and the probability that there is another identical one is almost impossible.

Even if you forget everything I have said so far, remember this similitude: metabolomics is like a barcode that contains an enormous amount of information. The metabolomic fingerprint is unique, just as a barcode is unique (Fig. 4.2). On a flight, every passenger has a boarding pass with a specific barcode; there is a single barcode on every product we buy at the supermarket. If some 50 of my readers sent me a urine sample, with metabolomics I could tell them one from another with no duplicates or errors: this is personalized medicine.

Figure 4.2 The metabolomic fingerprint (or profile) is like a barcode.

What is at the root of metabolomics? An intriguing discovery

How does metabolomics work? If we look at the first box in Fig. 4.3, we see a system in which the nodes (gray circles) are all equal and have few links. This is the so-called Gaussian system. But our organism's metabolism follows a different law, that of a scale-free network, which is shown in the lower box of the figure: not all the nodes are the same: the black ones have many links and are more important than the others. These are the ones that determine with precision the important characteristics of a structure, a disease, a patient. Every situation or disease presents a specific set of metabolites. This system of nodes creates articulated architectures that form the complex networks that are based on Leonhard Euler's graph theory, formulated by the Swiss mathematician in the first half of the 18th century. He applied this to the famous problem of the seven bridges of Königsberg.

How does metabolomics work? It goes in search of the black nodes (in this case the metabolites) that enable us to characterize a context univocally: I do not necessarily need all the information available to characterize the whole!

A brief example. In 2008, the Museum of Modern Art of New York acquired some works by Brendan Dawes as part of a project aimed at creating a visual "fingerprint" of entire films; the project began in 2004 and is entitled *Cinema Redux™: creating a visual fingerprint of an entire movie*. Each work is devoted to a film and is presented as a panel with a series of single images arranged in straight lines, one line under the other: each line represents a minute of the film reproduced in 60 frames, each one shot at a 1 s interval. Imagine that we approach one of these panels without knowing the title of the film: do we have to follow the entire sequence of frames to understand what we are seeing, or is it enough to catch certain details? We look at a frame and see a cliff. A few frames on we see Kim Novak and James Stewart on the top of the

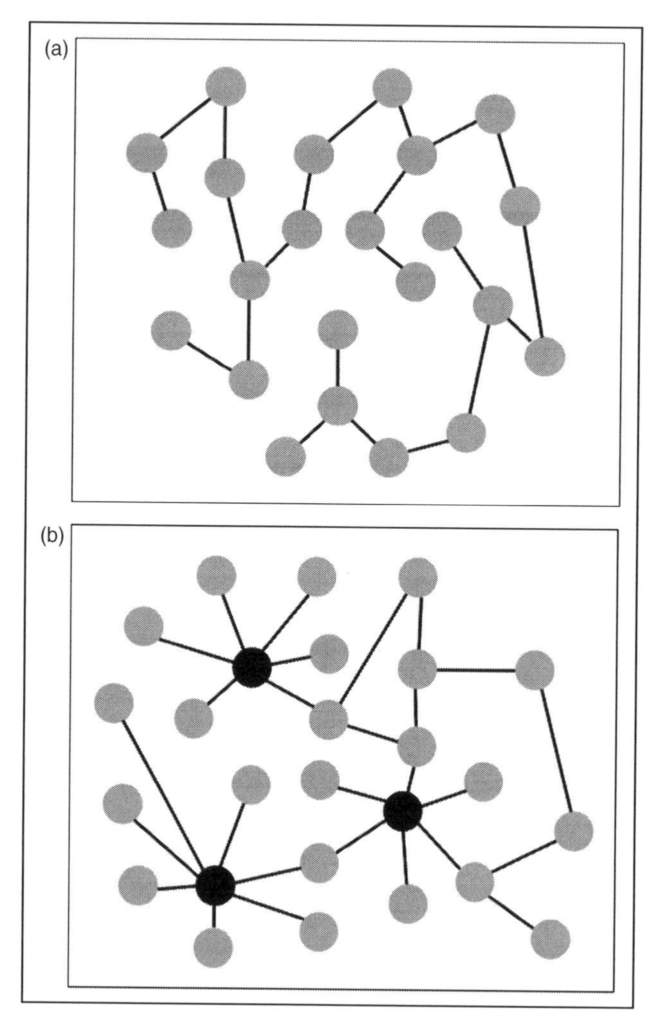

Figure 4.3 Different networks. (a) Example of a network with the Poisson (Gaussian) typology: the nodes (gray circles) are identical, each one with few links. (b) Example of a scale-free network: not all the nodes are the same. Some (the black ones) have a large number of links (hubs) and are much more important than the others.
Modified from Perpiñá Tordera (2009).

cliff: we don't have to continue to the end of the series, including the closing credits, to understand that the film is Alfred Hitchcock's *Vertigo*. In the same way, if I see the eyes of Bette Davis, I don't need to see the rest of her face to understand that it is her. If we observe a single detail of a painting, for example, a child with a red cap holding a loaf of bread and I am versed in the Flemish School, I can be sure that it is *The Wedding Feast in a Barn* by Bruegel the Elder.

Scale-free networks

Do you remember what I said about the organization of a cell being similar to that of one of our cities? In the same way I was quite intrigued and surprised to discover that the way our metabolism is organized is also the way in which other essential systems are organized, from our brains to the whole of our social relations.

Table 4.3 presents some examples of systems governed by the laws of scale-free networks, which range from the stock exchange to the writing of scientific articles, from our brains to the social networks, from the electricity grid to the telephone network of a large country. When we read a scientific article, the most important author (the most important node) is almost certainly the one who appears at the end of the list of authors, where by convention the leader of the research team appears. Thus, Gérard Depardieu and Catherine Deneuve are the black nodes of a film whereas the supporting roles are the gray nodes. On the stock market, the most important companies quoted pull the others up or down, so it is a scale-free network. It is at least singular that the law that governs our metabolism is the same as the one that regulates our social relations, the structure of the web, the network of roads, railroads, and hospitals. The stations of Rome, Milan, Naples, Bologna, and Florence are black nodes: this is a hub-and-spoke system. This is also the system that governs metabolomics: once again a scale-free network.

The organization of a system with the properties of a scale-free network makes the system more stable and capable of performing many tasks at the same time, just as occurs in our brains and metabolisms.

Another important concept is what in physics takes the name of the *accordion effect*. To explain it, here is another example: I am driving along the narrow Amalfi coast road in a sports car. In front of me is a bus. If the bus goes fast I can drive either

Table 4.3 Systems governed by the laws of scale-free networks

Scale-free networks	Nodes	Links
Cell metabolism	Metabolites	Participation in the same reactions
Brain	Six areas of great importance	The most important parts of the connectome
Scientific research	Scientists	Writing of scientific articles as coauthors
Hollywood	Actors	Participation in the same film
Stock market	The blue chip shares on the market	Stock market trend
Sexual relations	Persons	Sexual intercourse
World Wide Web	Social networks (Facebook, Twitter)	URLs (Uniform Resource Locators)
Highway system	Cities	Highway network
Railroad system	Stations	Railroad network
Hospital system	Hospitals	Hospital network

Modified from Perpiñá Tordera (2009).

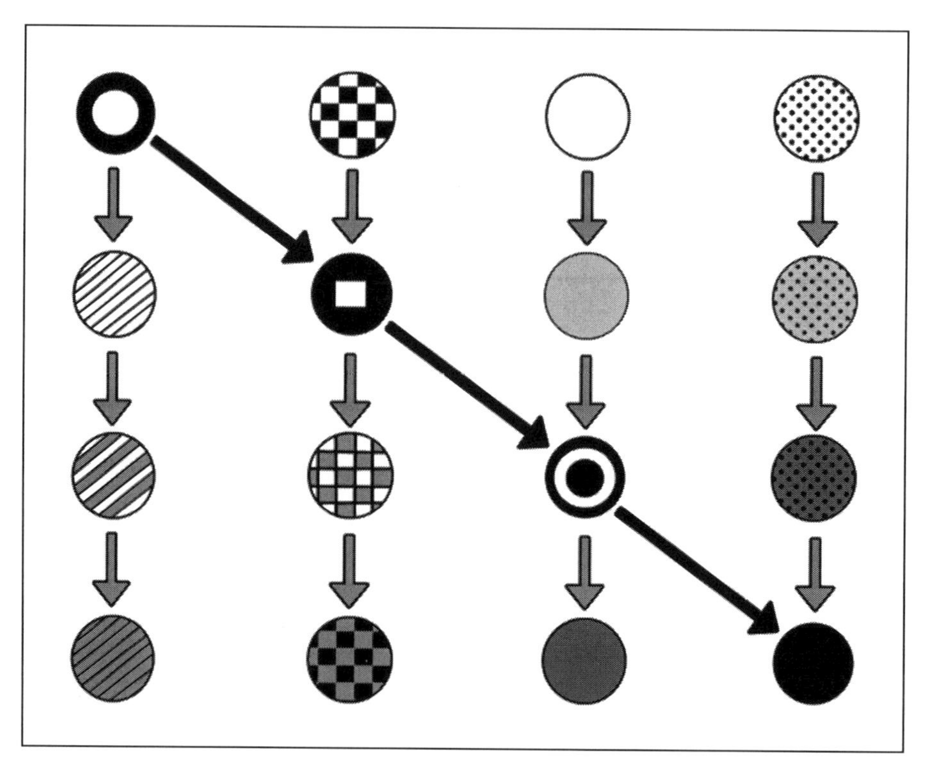

Figure 4.4 Another approach to metabolic processes. The metabolism has many interconnections between metabolites. The gray arrows represent four individual metabolic pathways as they are viewed in traditional biochemistry. The black arrows show the opportunities offered by metabolomics: there is a metabolic pathway that connects these four pathways by means of intermediaries for each one. From the functional standpoint this new pathway (highlighted in black) may be more important than the metabolic pathways as traditionally analyzed.
Modified from Kotze et al. (2013).

fast or slow, but if the bus goes slow I have no choice but to go slow. It's the bus that governs. This is true of a blue chip share on the stock market and of the most important metabolite of a disease.

Fig. 4.4 shows an alternative way to approach metabolic processes.

Network medicine

We owe to a physicist, Albert-László Barabási, the concept of network medicine, a different and more complex approach to medicine, perhaps the only one that can give us a picture of the extraordinary biological complexity of human beings. Barabási holds that the genome is not a medical panacea, but represents only a part of the problem to solve: it is only a partial list. He uses the metaphor of a mechanic who would not be able to repair a car if he did not know how the different parts interacted with the other components.

Network medicine is based on a series of advances in the field of network theory. The description of network medicine is quite complex and beyond the scope of this book, but I will briefly mention some of its essential elements. The networks that operate in biological, technological, or social systems are not random, but arranged on the basis of specific organizational principles, of which the hub-and-spoke system is fundamental. The node may be a gene, a protein, or a metabolite. The link (or edge) represents the interactions of the nodes of a network, for example, the ligand between protein and protein, or the connections between diseases based on a common genetic origin or shared phenotypical characteristics.

To try to understand and grasp the essence of the theory we must distinguish the topological, the functional, and the disease modules (Fig. 4.5).

A topological module represents a zone of the network thick with nodes and links, where the nodes tend to connect to nodes in the same zone rather than to those outside it. A functional module represents the aggregation of nodes having similar functions or which are correlated within one and the same zone, where the function assumes the role of a gene or one of its products, a protein or a metabolite, in defining recognizable phenotypes. Finally, a disease module represents a group of network components which together contribute to a cell function whose distribution results in a particular disease phenotype. These three concepts are evidently correlated. For example, the components of a topological module may have closely correlated functions and thus correspond to a functional module, and the disease may be the result of the inefficiency of one of these modules: that is, a functional module may also correspond to a disease module. Disease modules have three other characteristics: they are not necessarily identical to, but merely superimpose themselves on, topological and functional modules; every disease module refers to a single disease, that is, every disease has its own unique module; finally, a gene or a metabolite may be involved in many disease modules, which means that many of these modules may be superimposed. The gaining of knowledge and the characterization of the different disease modules will be important steps forward in network medicine. The European Union has launched the *Coordinating Action Systems Medicine* (CASyM) *project,* which aims to implement systems medicine.

Big data and... ten P medicine

Immunoistochemistry communicates with metabolomics in searching through thousands of data to find something (which Hoffmann has suggestively called "life's ratchet"). These data cannot be found using the classic criteria of anamnesis, physical and laboratory examinations, and imaging. Big data and information technology promise to change the world. To come to grips with the extreme biological complexity of our organism and each of our organs, the completeness of enormous amounts of data is of extraordinary worth if assessed holistically with the -omics disciplines. As

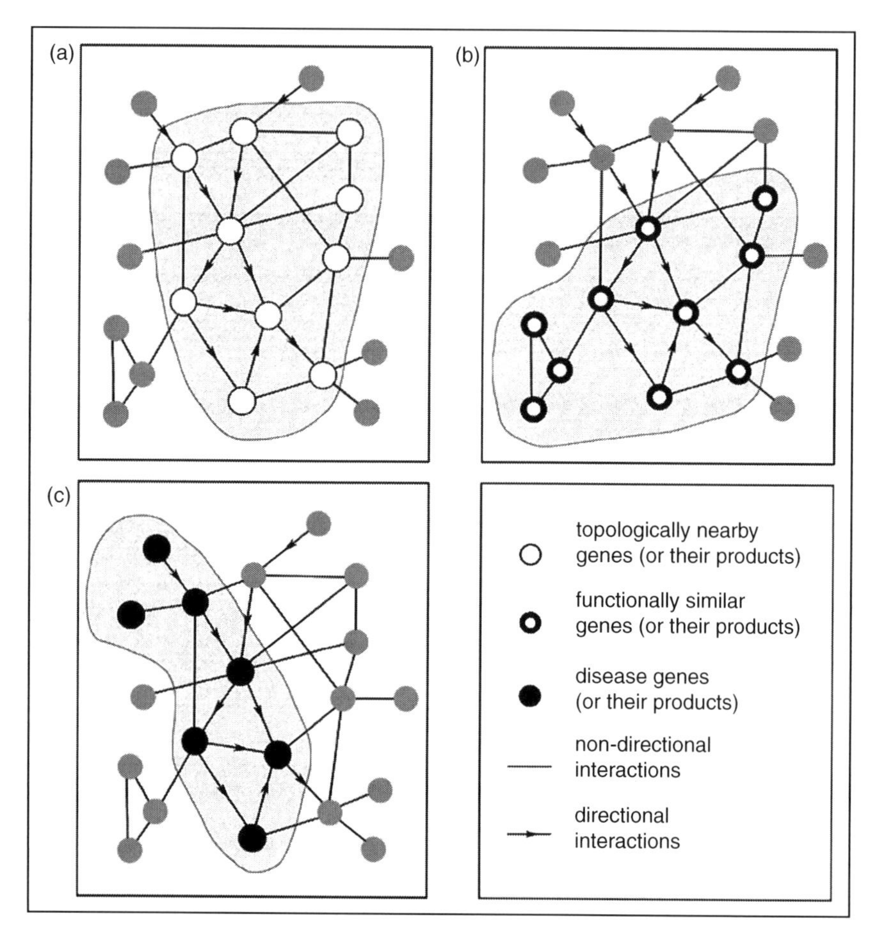

Figure 4.5 Topological, functional and disease modules. A topological module (a) represents a zone of the network dense with nodes and links where the nodes have a greater tendency to connect to those in the same zone rather than to those located outside of it. On the contrary, a functional module (b) represents the aggregation of nodes having a similar function or connected within one and the same zone, where the function assumes the role of a gene or one of its products, a protein or a metabolite, in defining recognizable phenotypes. Finally, a disease module (c) represents a group of network components which together contribute to a cell function whose destruction results in a particular disease phenotype. Modified from Barabási et al. (2011).

previously mentioned, Aristotle said that the whole does not correspond to the sum of its parts (I mention this for the third time on purpose). Today, if you want to understand the whole, we say that you must study it by means of systems biology, which offers the possibility of discovering our extraordinary interindividual variability. Today, we have huge amounts of data on hand and also large algorithms that enable us to solve a problem in times that are thousands of times shorter than the systems now commonly in use: this goes under the name of *data-driven intelligence*. Human capacity to process large data sets without outside help is comparable to that of a dog trying to solve arithmetic problems. There are problems within human reach and those within the reach of data-driven intelligence: metabolomics is extremely complex and belongs to the second group.

Futurologists foresee that shortly when a baby arrives at the emergency room a simple, ergonomic and cheap urine dipstick with preset metabolites will immediately orient the diagnosis toward the pathologies that must be treated quickly, or the baby will receive a shirt or bracelet with biosensors able to detect the main metabolites in its perspiration and then wirelessly transmit the information to the nurse's computer for an electronic triage.

We can also imagine how metabolomic data will be used in large general hospitals together with the customary laboratory tests. Physicians will examine them, cross-check enormous amounts of data and this will enable them to pinpoint the number of possible pathologies. Physicians will arrive at the diagnosis and begin personalized treatment much earlier than now, thus avoiding possible, even serious, collateral effects, such as irreversible liver or kidney insufficiency, on the basis of the patient's data even before deciding on the therapy. Only the right medicine for the right patient will be administered with no side effects: the clinician's dream.

I am often asked: "But will doctors no longer be doctors?" To answer, I recall that when the widespread use of echography was proposed the most conservative physicians had asked the same question. Today can we imagine being without that far more sensitive "third hand," the ultrasound probe, in the clinic? In the same way, big data offer a prodigious opportunity. What is extraordinary is not only the larger amount of data available, but also the fact that today with these data we can do something extremely useful and important.

Once again, an epochal change is taking place in the way we perform scientific experiments. Historically, an experiment was planned and it was decided what data to collect and process. Today, this system is under revision, facilitated by the drastic reduction in data storage costs in the last few decades. Researchers can now collect information on everything and then search for the important elements and associations in the data. Sometimes the associations are unexpected and can come only from the analysis of enormous amounts of data apparently far apart and unconnected. A

Table 4.4 Ten P medicine

• Personalized
• Perspective
• Predictive
• Preventive
• Precise
• Participatory
• Patient-centric
• Psycho-cognitive
• Postgenomic
• Public

striking example of this can be found even outside the medical world. The analysis of a huge amount of data on supermarket purchases has led to the discovery that on damp or rainy days customers buy about 30% less fruit and vegetables; thus managers are increasingly interested in weather forecasts and save a great deal of money by ordering less for such days.

If we apply big data in all fields of human activities, why not do so in medicine? This is the most important field (I admit that as a physician I have a conflict of interest in this) or at least one of the most important of human activities: public health accounts for approximately one-sixth of the total of public expenditures in the countries that apportion the most to social and health matters. Besides improving people's health, which is obviously the most important thing, there would be an enormous saving of money that could be set aside for further improvements, thus triggering a virtuous circle. An example of the use of big data in health care is the possibility of tracing the flow of epidemic events through data on cellphone activity: the highest concentration of use in given urban areas can give us a map of the probable diffusion of an infection. We can also think of the possibility of cross-checking all the data of our genome with those of our environment (the so-called *exposome*), in reasonable times and at affordable prices: a challenge for the immediate future.

In reality, these technologies and their application do not weaken the role of physicians: on the contrary, they represent a formidable instrument for extending their diagnostic potential and make possible 10 P medicine: personalized, perspective, predictive, preventive, precise, participatory, patient-centric, psycho-cognitive, post-genomic, public (Table 4.4).

The new technologies will enable us not to anticipate the future, but to make it possible.

Perinatal programming

Are the fetus and the newborn the parents of the adult?

Our life's journey from conception to birth, from intrauterine (prenatal) life to extrauterine (postnatal) life, through infancy to adulthood is a continuous journey during which our different organs and apparatuses progressively acquire a structure and a function.

As Jean R. Oliver wrote: "Structure does not determine Function or vice versa, but both are simply different ways of regarding and describing the same thing."

Biological age corresponds to the sum of gestational and chronological age (Fig. 5.1).

Up to not so long ago, we thought that the neonate was essentially a healthy being and that the adult deteriorated progressively and inevitably with the passing of time. Today it is thought that the neonate contains in itself, starting from before birth, a predisposition to become ill. A great deal is decided in the prenatal and perinatal periods, after and beyond genetic heredity: it is what we call epigenetics.

David Barker coined the expression "fetal and infant origins of adult disease," and with his studies he demonstrated the relationship between low neonatal weight and long-term development of cardiovascular diseases and glucose intolerance. This concept was later also extended to excesses in eating, diabetes during pregnancy, and stress other than nutritional. From the standpoint of the mechanisms responsible for this phenomenon, there is a lack of energy and oxygen. Besides a cascade of alterations in hormonal factors and epigenetic modulations that I will write about in the following chapters, this determines a mitochondrial activation of the fetus to make up for these deficiencies. Hyperactivity of the mitochondria, the sources of energy conversion and cell respiration, causes a boomerang of oxidative stress in many cells of the organism: pancreatic β-cells, hepatocytes, myocytes and, even before, those of the placenta.

Well then: are the fetus and neonate the parents of the adult and not the contrary? This is a provocative question that comes directly from the concepts that David Barker illustrated in the hypothesis that goes under his name and that develops the idea of perinatal programming. In agreement with his theory, not only is the neonate not a small adult, but the neonate (or even the fetus) may be the "father" of the adult since the response of a developing organism to a specific change that takes place in a critical time window in the perinatal period alters the trajectory of development from the qualitative and quantitative viewpoints, leading to permanent effects on its phenotype.

If we administer a certain dose of cortisone to pregnant rats prior to the 15th week or after the 18th week of gestation, practically nothing will occur in the neonate's cardiovascular apparatus. But if we administer the same amount of cortisone between

Metabolomics and Microbiomics. http://dx.doi.org/10.1016/B978-0-12-805305-8.00005-4

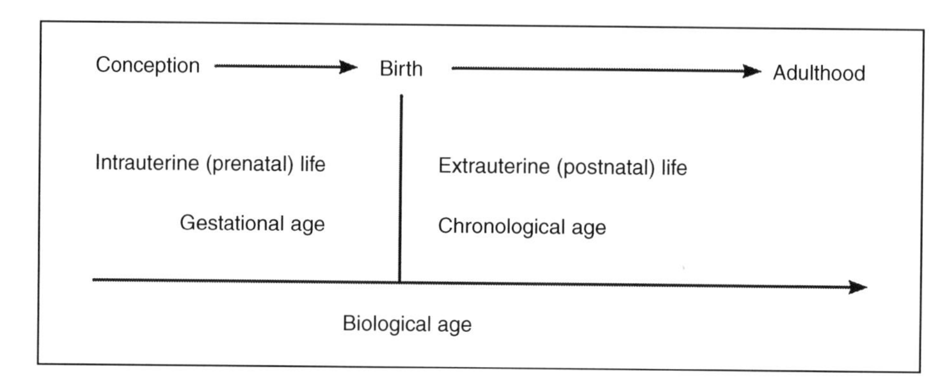

Figure 5.1 A journey through time: how a human being forms.

the 15th and 18th week (the critical time window) the offspring at the time of birth and then after will present extremely serious hypertension.

Each trimester of pregnancy features windows of vulnerability of specific organs and apparatuses, as shown in Table 5.1. One of the future challenges will be the transformation of these windows of vulnerability into ones of opportunity, with the performing of preventive or corrective actions.

Referring to the table, a problem that arises in the third trimester of pregnancy may cause neurological repercussions. David Barker, in an interview entitled *How the first nine months shape the rest of your life* published in *Time* in 2010, pointed out that even the major psychic problems such as depression, schizophrenia, antisocial personality, autism, and the attention deficit hyperactivity disorder, derive from that period of life. This is not surprising if we consider that most neurons and synapses form in the third

Table 5.1 Vulnerability windows of specific organs and apparatuses in the three trimesters of pregnancy

First trimester
Obesity Dyslipidemia Cardiovascular diseases Hypertension
Second trimester
Lung diseases Kidney diseases
Third trimester
Diabetes Depression Schizophrenia Antisocial personality Autism Attention deficit hyperactivity disorder

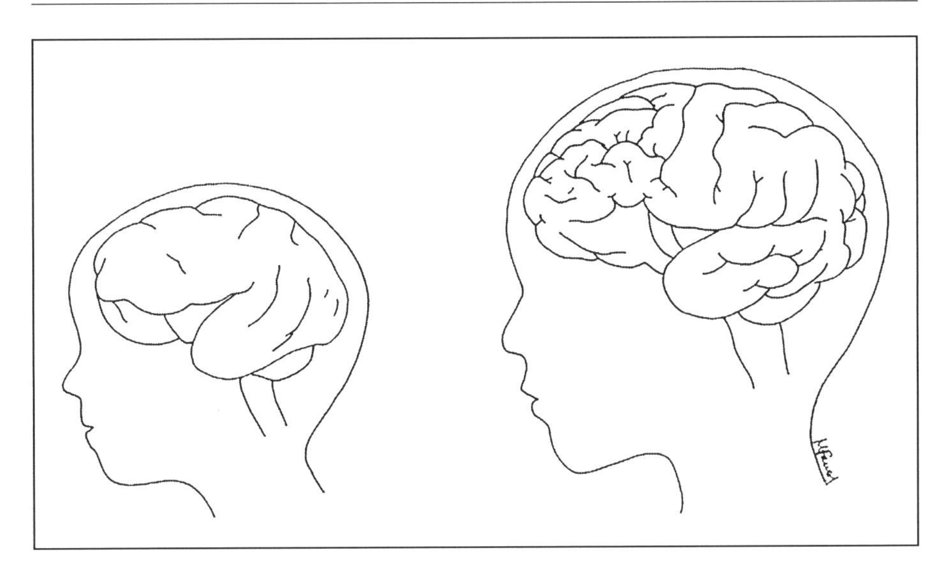

Figure 5.2 Differences between the brain of a neonate at 8 months and that of one born at term. At 35 weeks the brain weighs only two-thirds of what it will weigh at the end of gestation. At 34 weeks the cerebral cortex is only about half that of a neonate at the end of pregnancy.
Modified by Margherita Fanos from March of Dimes Foundation (2012).

trimester, since the brain is the most important organ it is the last to develop in the final period of gestation. At birth, the brain represents about one-tenth of body weight. The more primitive brain structures, the subcortical ones, mature before those of the cortex, which are more highly evolved. At 35 weeks, the brain weighs only two-thirds of what it will weigh at the end of gestation. At 34 weeks the cerebral cortex is only about half that of a neonate at the end of pregnancy. In Fig. 5.2 we see the differences between the brain of a neonate at 8 months of gestation and that of a term neonate. My question is: are we sufficiently aware of these important differences? In particular, it would be a good idea to emphasize to obstetricians that the choice to deliver a late preterm neonate with a gestational age between 34 and 36 weeks must be based on absolute necessity, otherwise one-third of its brain will mature in unforeseen, unusual, and unexpected environments, with not well-known short- and long-term consequences.

The womb is more important than the home

What we know today is that in all likelihood the most important period of an individual's life is the time spent in the womb. "The womb may be more important than the home" wrote Barker. During gestation, mother and fetus exchange signals, a sort of "biological SMSs," in an ongoing, intimate relationship, but we still do not know how and to what extent these mechanisms take place. Many are the questions that need to be explored and understood: in what conditions is the fetus safe and protected?; how

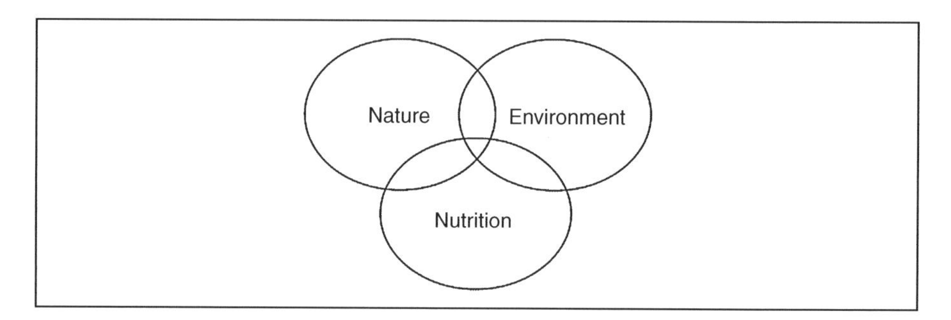

Figure 5.3 Nature (or genetics) and environment (or epigenetics) (*nature and nurture*). Of the epigenetic factors, nutrition is so important in the child's growth period that it has a circle of its own.

would it react to medicines administered to the mother?; what will it absorb from the mother's diet?; will it be born in a situation of over- or undernutrition?; what are the conditions in what we can consider an appropriate environmental context? These are but a few of the queries from which to start, because in reality fetal life is an area of research that is still for the most part virgin territory.

In general terms, the first thousand days from conception are decisive: although chronological life begins with birth, they are days of biological life: nutrition in this period plays a strategic role. In fact, if genetics plays an important role, one that is just as important, if not more so, is played by the environment (nature and nurture).

Wondering if nature or environment is more important is like asking oneself if in a rectangle the length or the width is more important, as the psychologist Donald Hebb stated. In any case, I agree with Wollheim: "a good environment is not a luxury, it is a necessity". In the environmental sphere, nutrition is so important that it plays an autonomous role and it must be emphasized that both hypo- (too little) and hypernutrition (too much) are forms of malnutrition (Fig. 5.3). The practical corollary to this statement is that the role of the obstetrician, the neonatologist, and the pediatrician is, and will continue to be, more and more decisive for the future of an individual even in adulthood in terms of prevention, early diagnosis, nutrition, and personalized treatment.

Starting from the third trimester of pregnancy, most of the cerebral mass and connections form, the function of the immune system stabilizes, and the intestinal microbiota is consolidated. As we will see, these three situations, which at first glance are separate, are in reality all functionally connected for the integrated maturation of the entire system of the organism.

Return to the future: the placenta and perinatal programming

David Barker, focusing his attention on the importance of the fetus and what occurs during life within the womb, has brought about an authentic revolution in medical thought. The understanding that our destiny as concerns health or illness is for the most part decided and laid out in our mother's womb, which is the basis of the concept of perinatal programming, is becoming increasingly widespread in the scientific community. The number of studies that emphasize the extraordinary role played by epigenetic mechanisms and their impact on the future life of each individual is growing day after day.

From these studies emerges the fact that not only women are involved, as we will see further on, but also men: for example, undernutrition during intrauterine life in a critical time window of a male fetus leads to perturbations in sperm in adulthood by turning off the activity of certain genes through a mechanism known a methylation, as I mentioned previously; this effect is of a transgenerational nature and is handed down to following generations.

Researchers are also recognizing that to understand epigenetic mechanisms in depth they must move the cursor farther and farther back in time and position it at the placental level. Although it is the life support of the fetus, the placenta has amazingly been for the most part neglected in their studies and we still know too little about this organ. Only recently have scientists begun to suspect that the placenta is not sterile as was thought in the past, but instead possesses, even in normal conditions, a microbiota of its own, which is to say a small population of "good" bacteria that shapes the fetus' immune system. It is worthy of note that the placental microbiota is closely correlated with that of the mother's oral cavity (tongue, tonsils, saliva and, to a lesser extent, the supra- and subgingival plaques and the throat), while it is completely different from that of other parts of her body (skin, nostrils, vagina, and intestine).

Perinatal programming significantly influences many organs and tissues such as the brain, heart, adrenal gland, kidney, lung and liver during their development and maturation and, in final analysis, their susceptibility to disease. In the near future, research will also reveal its impact on still other organs. Perinatal programming of the brain, which has led to the hypothesis of its possible connection with neurodegenerative disorders in the adult, such as Alzheimer's and Parkinson's disease, is a case in point. Once again, not only genetics is important, but so is epigenetics: everything appears to take place on the basis of a genetic predisposition on which intrauterine epigenetics is superimposed. This is why genetics does not necessarily decide our destinies. Genomics succeeds in forecasting some simple features of our organism, such as blood type, the type of cerumen (squamous or fluid), or alcohol tolerance, but it fails to predict height, color of eyes or hair, or lactose intolerance: the latter are in fact complex elements determined by the decisive contribution of epigenetic factors.

The mysterious trees of life

The organism has little imagination, even though in appearance this is untrue. What changes extraordinarily is the scenario of the pathological representation in the different organs and apparatuses, but the mechanism at the base of the condition of disease is quite simple and can be expressed in two words: cell hypodysplasia. This means that there are fewer cells that function less following a reduction of the arborization of various structures inside the organs (e.g., cerebral connectome, bronchial tree, ureteric bud, and so on), accompanied by a reduction in the branching of the vascular tree. Taking our cue from Denise Grady's fascinating title, we can call these structures "the mysterious trees of life." If someone has said that in a tree and its hundreds of thousands of leaves there must be the hand of God, this statement is even more effective if we observe the extreme complexity of the structures that make up the human organism and the complex trees working in the different organs.

The placenta is the first of these trees: when it is strongly involved in pathologies, especially before the 26th week of pregnancy, it probably has a negative influence on the formation of all the other trees. One example will suffice in this case. It is thought that certain psychic disorders of the child and young adult have their origins in exposure during fetal life to negative factors, among which are hypoxia and reoxygenation. A recent study demonstrates that the placenta, in response to altered oxygen concentrations, is capable of releasing substances that may damage the developing neurons, at least in experimental conditions. These data appear to confirm what some authors, such as Rees and co-workers, have highlighted: fetal brain damage may occur not only because the amount of oxygen transported to the brain is not enough, but also due to the accumulation in the fetal circulation of reactive substances released by the placenta that negatively influence the brain's vascularization and metabolism. A placental dysfunction is central to many complications in pregnancy such as preeclampsia, a serious pathology for both mother and fetus, especially in terms of mortality, restricted intrauterine growth, and death in the uterus. The actual molecular physiopathology of placental dysfunction in these conditions is unknown: perhaps only the new -omics disciplines will allow us to advance in this field. In any case, if today we asked ten obstetricians if the placenta is an innocent bystander or an active participant in what takes place during fetal life, nine of them would instinctively choose the second answer. In sum, the placenta is the "flight recorder" of intrauterine life.

Journey to the center of the uterus: "I was born to replicate myself"

To represent the importance of epigenetics and nutrition during fetal life, I will give an example concerning intrauterine growth restriction. Let us imagine that first we are fetuses in the womb and then neonates.

I am an apparently healthy baby at the 50th percentile by weight, length, and cranial circumference, which means that I am a perfect specimen of the average population. My mother goes to the obstetrician regularly for checkups, but unfortunately they find I am moving away from my weight percentile (from the track of my genetic potentialities). At the time of a checkup I am at the 30th percentile, which means that 30 fetuses of my sex weigh the same or less and 70 weigh more than I do at this stage of gestation. At a later checkup I am at the 8th percentile (8 fetuses of my sex weigh the same or less and 92 weigh more in this phase of gestation). This means that my weight is not increasing the way it should. The obstetrician is worried, mother and father even more so: what is happening to me?

For now, we can skip the search for the causes and concentrate on the consequences, since intrauterine growth restriction is linked to at least four events that have not yet been assessed as concerns their impacts and short- and long-term implications.

The first effect to be taken into account is that in the immediate future my organism will become more resistant to further negative changes that may involve me and I will assume what is called a "thrifty phenotype." The refrain of an Italian song, *"Come si cambia, per non morire..."* (How we change, not to die...), is confirmation, as if there was a need for it, of Darwin's lesson on evolution: to survive we must change. It is as if after a negative experience my organism has erected a curtain wall even higher and more difficult to penetrate for the next enemies or adversities I will have to face.

The second point is that what happens to me takes place not only to give me the chance to escape from the hands of the intensive care neonatologists, but also to become "eternal." Neruda's famous expression, "I was born to be born" is not true in this field: we have to change it to: "I was born to replicate myself." The aim of everything that is happening to me, and the changes in me, is to give me the possibility of replicating myself and, in a certain sense, of becoming "eternal." I will be small and have a high probability of dying or having problems, so to increase my chances of survival at birth I'll try to protect my brain and heart at the expense of the other organs, especially the intestine, the organ that will have to make the biggest sacrifice, the liver, the kidney, and the growth of my body itself. If I want to survive I'll have to change, with important modifications that will enable me to reach adolescence and sexual maturity (at least 12–14 years). But there is always the down side. To reach this goal, assuming I survive the cause of my intrauterine growth restriction, I'll have to change my initial metabolic course and pay for this in the long run with metabolic syndrome, the mixture of clinical conditions that includes diabetes, hyperinsulinism, chronic kidney insufficiency, high blood pressure, a propensity to develop cardiovascular diseases, too much cholesterol, and so on.

The third fact that occurs is probably the most frightening: epigenetically, I will hand down to my offspring what is happening to me! My genome is still the same, but the epigenetic modifications will be transmitted to two or three later generations. Limiting studies to genomics alone appears now to be part of archeological medicine: we need to arrive at a better understanding of the interaction between genome and environment.

The fourth consequence is the most important from the practical standpoint. Given that what has happened to me has changed me forever, I'll have to be very careful about what I eat and how I'm treated after my birth. A metaphor will explain this better: I was destined to become a car with a medium-sized engine, but now, despite my efforts, I'll be a car with a small engine and so I'll have to use the right kind of fuel for my new situation. Instead, pediatricians, and in particular neonatologists, following today's guidelines, decide to feed me as if I were still in the womb and had to keep up with the growth rate of a fetus. They feed me what today is considered the best food for my brain, but which is aggressive for the rest of my organism: it is as if they filled my tank with Formula 1 fuel, while I have the characteristics of a city car. This discrepancy, this mismatching between what I am fed and the overall requirements of my organism may anticipate what I would prefer to postpone as long as possible: the onset of metabolic syndrome.

Metabolic syndrome

Metabolic syndrome occurs most frequently when diet at birth or during infancy differs significantly from diet during intrauterine development. For example, growth too soon and too quick may cause problems in the long run. The same can be said for excessive growth. I will try to illustrate briefly some important concepts with three slogans.

- Grow now, pay later.
- Live fast, die young.
- Bigger is not always better.

Appropriate nutrition in the early periods of life favors improvement in the functional development of the different organs, especially the brain and the immune system, with a reduction in the risk of diseases, both transmittable, such as infectious diseases, and nontransmittable, such as diabetes, obesity and their association, known as diabesity. Inappropriate nutrition associates with negative outcomes: an excessive weight increase in the first year of life or the early periods of infancy is the strongest element predictive of the percentage of fat in adulthood.

My first professor of pediatrics, Dino Gaburro, a great nutritionist at the Medical School of the University of Padua, the school that taught us how cow's milk should be diluted and fortified for administration to neonates and the unweaned, and later at Verona, repeated that a child's growth is a march and not a run. If I want to make the pilgrimage to Santiago de Compostela there is no need for me to run the first 3 days and wind up half-dead on a bed in intensive care and with blisters on my feet. The sensible thing to do is plan a series of stages that are fatiguing but not impossible to walk.

Finally, if we were to sum up in a few words how to prevent metabolic syndrome, or at least limit its consequences, we could say that a careful diet is based on two fundamental points: use mother's milk (we will never stop emphasizing its uniqueness and extreme importance) and avoid periods of excessive weight increases. This is what my second professor of pediatrics, Luciano Tatò, a great endocrinologist, he too first at the School of Padua and later of Verona, would remind me of.

As concerns members of the family, we know in all fields, including diet, that the only thing that matters for babies and children is the example: *do as I do* and not *do as I say.*

What do we need to know about perinatal planning? Ten golden rules

I will try to sum up important concepts in a few points, some mentioned or discussed at length previously and some still to be addressed. Beside each of these points I have used a motto, a slogan (a word from Scots Gaelic that means "battle cry," in our case "concept cry," as if it were printed in boldface, underlined, and with an exclamation point). I often use slogans because they are easy to remember, they better penetrate individual defenses and often also bear a "supplement of soul" to use an expression dear to Bergson. And now, the ten golden rules.

1. *Transform windows of vulnerability into windows of opportunity*: there are certain critical periods during development in the womb, in which suboptimal conditions expose the fetus to risk. These periods differ for the different organs. Knowing these things, we can intervene by monitoring the nutrition and wellbeing of the expectant mother and consequently the health of the fetus.
2. "*In my beginning is my end*" (T.S. Eliot): perinatal programming has permanent effects on the organism in the short and long run and can modify susceptibility to disease.
3. "*The chief function of the body is to carry the brain around.*" (Th. Edison): perinatal programming determines the permanent structure of important organs, trying in all cases to protect the brain and then, in decreasing importance, heart, adrenal gland, kidney, and all the other organs.
4. *The placenta is not an "innocent bystander."* The placenta plays a key role: it is the extremely important organ of mediation between mother and fetus, and many of its aspects are still scarcely understood.
5. *I was born to breed*: compensation for changes in the metabolic trajectory when prenatal problems arise is not without costs; it is like a tax to pay to reach the capacity to breed and, with this, become "immortal."
6. *All that glitters is not gold*: the gold standard of today may show its limits tomorrow. Today it is thought that the gold standard consists of promoting the growth of the preterm neonate as if it were still in the womb, but this may turn out to be a mistake, at least for some babies. In fact, attempts to correct and reverse the effects of perinatal programming by following international guidelines may lead to unwanted results, as has been seen in the case of a too-aggressive diet for the single neonate.
7. *The fetus is not a small, highly preterm neonate*: the fetus reacts differently to suboptimal conditions than does the neonate and later the adult. In the same way, the child is not a small adult, the preterm neonate is not a small neonate, the highly preterm neonate is not a small preterm neonate.
8. *We are all different*: there is an extraordinary interindividual variability from one fetus to another. And they react to the same stimuli in different ways.
9. *You have to choose your parents, grandparents, and great-grandparents carefully*: the effects of perinatal programming are transgenerational. Once the saying was simply paradoxical: "You have to choose your parents." Today it is more appropriate to think in terms of epigenetic cycles of about a hundred years: we are what our great-grandparents were and we will influence later generations down to our great-grandchildren.
10. *Perinatal programming has different effects on males and females*: we must not confuse the social realm with biological reality. If we wish to offer a real service to both males and females we must commit ourselves seriously and without prejudice to the exploration of gender medicine, of which I will speak in greater detail later.

Our plasticity is impressive

The ability of a certain genotype to produce different phenotypes in response to different environments is defined as plasticity and represents part of an organism's adaptation to its surroundings. Our impressive plasticity descends from an inextricable interpenetration between genes and environment. Plasticity has high energy costs and is at its maximum in the period of neonatal and infantile development.

An example of plasticity is represented by the brain of the preterm neonate. The earlier the prematurity, the more its elasticity. In a neonate with a cerebral hemorrhage or asphyxia, the cells near the damaged or dead ones take over the functions of the latter. The results of these injuries are not necessarily as destructive as they are in the adult: a neonate's cells are like young, just-hired workers who can change jobs with a certain ductility and learn new tasks easily. The situation is different for older workers about to retire who have done the same job for many years. Their capacity to adapt to a new job is limited and they are inevitably slow in learning, compared to a much younger worker. It is much harder for them to find another job.

Plasticity in the planning of development has evolved so as to ensure the best opportunities for success of an organism that has to cope with different environments. The window of plastic development extends with different apertures from conception to adolescence and can be transmitted in a transgenerational way from one generation to the next and those that follow. What occurs in early periods of life can strongly influence human biology in terms of the growth of the neonate, the maturation of the child, the long-term health of the adult, and the elderly. In other words, events that take place early in life can shape our futures for better or worse. Nutrition in this scenario is of the utmost importance.

Nutrition in pregnancy

It has been demonstrated that women who regularly follow the Mediterranean diet at or around the period of conception run a lower risk of having malformed or growth-restricted fetuses than those whose diet is different.

A Norwegian study confirms that a diet with proper amounts of fruit and vegetables, fish, and abundant water intake significantly reduces the risk of preterm delivery. I mentioned earlier that Thomas Edison, the inventor of the light bulb, predicted that future physicians would treat and prevent diseases through diet. It is obvious that the quality of food is extremely important and the levels of environmental pollution with toxic substances must be taken into consideration.

Recent studies have focused on these elements and have found that the main environmental factor associated with preterm delivery is high concentrations in the urine of bisphenol A (BPA), a compound whose action may mimic that of estrogens and increase the probability of a preterm birth. BPA is present in most plastics that have now been in use for over half a century. This substance may also come from canned foods and beverages. Epigenetically, BPA causes alterations in the cerebral synaptic connections and these lead mothers to be less caring and the offspring to show more anxiety. As for the hormonal effects of BPA, here are two examples: the first comes from laboratory tests that have shown that female mice can distinguish males treated with BPA, which are less fecund, and are less attracted by them. The second example is from real life and is in the low number of spermatozoa in Chinese workers in factories that produce plastics and are exposed to BPA. Beware: BPA is among us and in us! Over 90% of us have BPA levels that can be traced in the urine. The use of BPA has now been banned in the production of baby bottles in the European Union.

Transgenerational epigenetic inheritance

It is thought that in the last 50 years the numbers and quality of spermatozoa have decreased to half the previous situation: a true disaster! Besides the negative influences of the environment, we know that two maternal factors, obesity and smoking, have an unquestionable influence on their sons' low number of spermatozoa. The determination of the fetus' sex and the development of testicles and ovaries take place at approximately 6–12 weeks and orient the male and female archetypes for development of the brain. The future mother's diet and health are important in the periconceptional period, but what is emerging strongly is that just as important is the mother's intrauterine experience, that is, her experience as a fetus in the grandmother's womb. Call it if you like, "transgenerational epigenetic inheritance." The mother's oocytes, from which today's children are born, were formed in the grandmother's womb. Thus we can say that the diet and health of a woman in the periconceptional period influence not only the gestation and birth of her children, but also that of her grandchildren, as we saw previously.

Early nutrition and the destiny of the individual

Early nutrition has profound effects on the organism. As an example, among bees it is only the difference in diet that leads a larva to become the queen.

Today we are living in a world in which a misalignment in evolution has taken place. To explain better, we are living in a reality in which our way of eating and our energy expenditure are quite different from the context in which humans evolved. Moreover, we live much longer than in the past and so there is no selective pressure against late-onset diseases in our lives. It is certainly unrealistic to imagine a return to the lifestyles of the past, but much can be done in any case. We have already said and repeated that prevention of the global epidemic of noncontagious diseases, such as diabetes and obesity, begins in the uterus. Control of *salt*, *sugar*, *spirits*, *saturated fats*, *starvation*, *sedentary life*, and *smoking* (both active and passive), the seven "S's", is extremely important. In conclusion, the fight to prevent many diseases begins during intrauterine life.

How we age depends on our fetal lives

From the cover of the *National Geographic* of May 2013, a child looks at us and the title states: "This baby will live to be 120."

There are still many questions surrounding aging that have not been answered fully. Why do some regions have many centenarians? What characterizes the healthy elderly, the "wellderly"? Some 80-year-olds do not present chronic diseases such as hypertension, heart disease, or diabetes, nor have they ever taken medicines to prevent or cure them: does it depend on their genes, their environment, or on both of these factors? Are longevity (life longer than the average of an organism) and aging (the progressive biological and physiological decline of an organism) the two sides of the same coin? Aging is negative for the single individual but advantageous for the species: the death of older individuals, no longer able to breed and therefore biologically useless, leaves room for the new arrivals, the young people able to breed: so, is aging written in our DNA?

Our pattern of aging is presumably decided before our birth. Even in this field the genes probably cannot explain everything: it is thought that they represent only 25% in determining longevity. A quite recent study went in search of the so-called "Methuselah genes," possibly present in the DNA of those over a hundred. In this case the entire genome of those over 110 years of age was examined. The researchers thought they would find the secret of longevity, but contrary to their expectations, they found no inside track for living a long life. There are no exceptional and protective genes: these extreme performances come from the combined and fortuitous association of a given genome with favorable environmental situations, such as a healthy climate, proper lifestyles, and diet: there are no "good" genes or "bad" genes, only certain genes for a certain person at a given moment (Fig. 5.4).

The genetic marking of stem cells from the umbilical cords of babies of low weight for their gestational age (too small) and overweight for gestational age (too big) differs from that of babies of appropriate weight. The fetus is the father of the adult: this is the paradox!

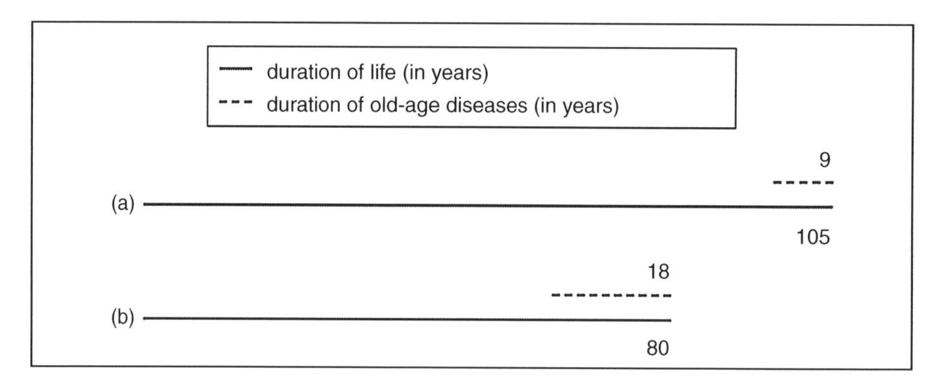

Figure 5.4 Centenarians (a) reach the blowing out of a hundred and more candles because they are healthier owing to a good combination of genetics and epigenetics. In persons with an average lifespan (b), diseases connected with old age (heart diseases, stroke, diabetes, senility, cancer) begin earlier and last longer: the figure represents them with dashed lines. Modified from Hall (2013).

The mTOR protein: a two-faced angel

I wish to speak of the mammalian target of rapamycin (mTOR) protein. Although we must not fall into the trap of entrusting all our desires to genes or proteins, the mTOR protein has an interesting story to tell, that of a molecule that looks like a two-faced Janus, one positive and the other negative, but out of phase in time.

This protein plays an essential role in prenatal life and the first periods after birth (it is present in the process of nephrogenesis, for example) and enables the organism to grow by its presence in many metabolisms, especially that of insulin and glucose. So it is a true "guardian angel" for the child, both inside and outside the womb. However, after performing its task diligently, the mTOR protein deactivates and remains quiescent for years, only to be reactivated at a certain point and become an "exterminating angel," aging us and stimulating the onset of tumors. The mTOR protein acts as a food biosensor: when food is abundant its numbers increase, when food is scarce its numbers decrease. Its function is closely connected to insulin. It also reacts to other situations of cell stress, such as hypoxia and cell damage. The drugs that block mTOR, such as metformin and rapamycin, also block aging in animals and lengthen their lives.

Rapamycin connects the mTOR protein to Easter Island. While building the island's airport in 1964, many samples of plants, animals, and soil were taken. The dust taken from the ground contained an antifungal substance capable of lengthening the life of different animals. The substance took the name of "rapamycin" from Easter Island, which in the native language is Rapa Nui. Rapamycin thus acts as a blocker of the mTOR protein. In mice treated with rapamycin, the average lifespan increases by 14% in females and 9% in males.

Neonatomics and childomics

It is obvious that application of the -omics technologies is important in all studies concerning pregnancy and the first years of life, since these periods have a bearing on our entire lives, our state of health and illness, and our longevity. Together with

Table 5.2 The main fields of metabolomic studies in neonatology and pediatrics performed by the Cagliari metabolomics team

- Gestational and postnatal age, type of delivery, twin births
- Perinatal asphyxia, brain damage, and hypothermia
- Intrauterine growth restriction and children of a diabetic mother
- Acute and chronic neonatal respiratory pathology, bronchiolitis
- Cardiovascular diseases, persistence of the ductus Botalli
- Kidney pathology and kidney damage caused by medicines
- Sepsis and necrotizing enterocolitis, congenital cytomegalovirus infection
- Pharmacometabolomics and nutrimetabolomics (including the study of mother's milk)
- Autism spectrum disorders and pediatric aggressiveness
- Long-term diseases (of the adult) in those born with very low weight (e.g., cardiovascular diseases, autoimmune diseases such as scleroderma)

Modified from Fanos et al. (2013d).

Giuseppe Buonocore of the University of Siena and Michele Mussap of Genoa, I wrote an editorial in which we coined two neologisms: *neonatomics* and *childomics* (that is, application of the -omics technologies to the study of the neonate and the child). I think that in consideration of what I have written in this chapter on perinatal programming, these technologies may represent the right road to the future. In fact, an understanding of the physiological and pathological mechanisms that intervene in the first periods of life, both inside and outside the womb, will enable us to better understand and prevent many diseases in adulthood that were previously latent. In this sense, in our eight years of metabolomic research, the Cagliari metabolomics team has studied in detail, together with many other research teams, a large number of issues in the fields of perinatology and pediatrics, as shown in Table 5.2. I am also working with Corrado Moretti of the Sapienza University of Rome, in the search for practical, bedside diagnostic instruments to arrive at an early diagnosis of the main pathologies responsible for neonatal mortality.

From birth to the miracle of mother's milk

Being born: an adrenaline-charged action movie

Adrenaline is the stress hormone, but also one of risk and adventure. The adjective "adrenaline-charged" is often used in movie trailers to show that the film is an action movie with a fast and gripping rhythm. Imagine yourself at the center of a scene: you are on a plane with a parachute strapped to your back. At a certain point your captain says: "It's time!" You've got to bail out. How much adrenaline goes into your bloodstream? Actually, it is much less than the amount in the blood of a baby at birth with a perfectly normal delivery: the baby begins to produce extremely high levels of adrenaline and noradrenaline, stimulated by the compression of its head in the birth canal. If the birth is a difficult one, noradrenaline increases by over two and a half times compared to a baby delivered with a physiological birth. In relative terms, a baby can have up to 100 times the noradrenaline of an adult at rest, 50 times that of an adult after a sauna, and 4 times that of an adult doing strenuous work. The neonate is more stressed than a marathon runner, but it does not appear to be stressed at the time of birth during a normal delivery: stress seems to be good for it! The task of stress hormones is to goad the organism: it prepares us to face difficulties so as not to succumb: *fight or flight*. The trauma of birth causes a strong increase in adrenaline and noradrenaline not only in the bloodstream, but also in the bluish area of the brain (the *locus coeruleus*). There is a cascade reaction on the freeing of other hormones and chemical mediators (such as dopamine) that contribute to keeping the body temperature and glycemia stable, thus promoting and facilitating breathing. The intrauterine environment is relatively hypoxic compared to that outside the uterus, which is relatively hyperoxic. Being born is like going from the top of Mount Blanc at over 4000 m above sea level down to the sea in a short time: the neonate is like a skydiver and stress hormones make this extreme descent possible.

Delivery with a planned Caesarean section without passing through the birth canal is much less stressful, but this means that fewer stress hormones are produced and the neonate's conditions are different from those of a baby born through the vagina. New studies are concentrating on these differences and their short- and long-term consequences. I will speak of this later when I examine the microbiota of mother's milk.

Why does delivery begin?

Metabolomic studies suggest that the mechanisms that trigger labor have elements in common with preterm delivery and preeclampsia, a most serious obstetric complication. Today, nobody in the scientific community can explain exactly why the organism

Metabolomics and Microbiomics. http://dx.doi.org/10.1016/B978-0-12-805305-8.00006-6

presses the button of no return and labor begins. There is probably once again a deficit in energy, which is no longer sufficient to govern the maternal immune system and control and tolerate the "foreign body" represented by the fetus. At this point the hypothalamic stress axis is activated: the ACTH hormone stimulates the adrenal gland to produce cortisol, another hormone. This situation reduces the effect of placental progesterone and increases the production of estrogens, thus making the uterus more sensitive to the effects of oxytocin. Delivery is thought to start when the fetus' glycemia begins to decrease and/or it starts to consume about 15% of the mother's energy. This point is reached earlier by twins, and this is one of the reasons why twins are born preterm.

An understanding of these mechanisms is extremely important. Preterm delivery is the main cause of perinatal morbidity and mortality. Every year there are about 15 million preterm deliveries in the world, with enormous costs for society: in the United States alone the expenditure is 26 billion dollars a year.

So it may be that it is the fetus, and not the mother, that triggers labor because it is not getting enough to eat! In countries where periods of fasting are imposed by religious beliefs, the number of births increases repeatedly. When delivery begins, it is very hard to stop it: it is like a landslide.

The first meal is like the first breath (a hymn to mother's milk)

The first instants of life and nutrition at that time are of the utmost importance for the destiny of the newborn baby. Normally, the importance of its first moments is acknowledged for the lungs and the first breath, but in all likelihood this should be extended to the other organs as well. One of my professors at the post-graduate school of pediatrics in Verona, Giuseppe Zoppi, coined the phrase: "The first meal is like the first breath."

The best time for preventing diseases through diet is after birth. The dyad of the mother breastfeeding her newborn child is a unique model for understanding the importance of diet, both in terms of prevention and protection.

I won't go into detail about what we know of mother's milk, but here I do want to say that breastfeeding is the norm and the model of reference: all alternative methods of feeding must be measured against this model in terms of growth, health, development, and short- and long-term results.

Mother's milk is the first choice for babies' nutrition, whether they were born at term or preterm, according to the guidelines set by the American Academy of Pediatrics (AAP) and the World Health Organization (WHO). Both organizations state that breastfeeding alone is to be continued for 6 months. Following this, supplementary foods should be introduced while continuing to breastfeed up to 1 year or even more, depending on the wishes of mother and child, for the AAP and up to 2 years or more for the WHO.

Breastfeeding provides benefits for the health of the child in the short and long runs. In the early period it protects against respiratory and gastrointestinal infections as well as reducing the incidence of allergies, such as atopic dermatitis and asthma; in the long run it reduces the risk of obesity, diabetes of types 1 and 2, the coeliac disease, chronic intestinal diseases and boosts neurobehavioral development.

The growth of the intestinal cells

From Elsie Widdowson's studies, we know that the duodenum of breast-fed piglets gains 42% of its initial weight in the first 24 hours of life. This emphasizes the paramount importance of breast milk, when it is available, in the first moments of life. It is like the ideal software in the hardware for which it was created: it influences the rest of a person's life, triggering the entire cascade of the organism's metabolic processes. The epithelial mammary cell is considered a bioreactor for complex structures and bioengineering activities that interact fully in all of a baby's physiological processes: growth and development, microanatomy and function, physiology and metabolism, colonization and maturation of the gut microbiota, immunological orientation and consolidation, structure and organization of the brain.

It is hard to imagine anything more individualized than mother's milk: it changes while breastfeeding, from colostrum to transition milk to mature milk, during the day with morning milk more invigorating than evening milk, and finally during the feed, with an increase at the end of the meal in the fats that saturate the neonate's centers of satiety.

Microbiota of mother's milk

Up to little more than 10 years ago it was thought that mother's milk was without bacterial colonization and this was considered one of the advantages of breastfeeding. Today we know that it contains a certain amount of bacteria (microbiota) belonging to 50 genera and 200 species that come from sources inside and outside the breast. We speak of MOM (milk-oriented microbiota). The composition of mother's milk microbiota is highly individualized and unique for each mother; it has important implications for the colonization of the intestine and the short- and long-term health of the child. What we know is that these consequences on human health begin from the very first moments of life and follow a person's evolution in a triangular relationship: the guest (the neonate), mother's milk, and MOM. Just as the placenta has a microbiota quite similar to that of the mother's tongue and throat, the microbiota of her milk is quite similar to her intestine. If the mother has a dysbiosis (an alteration of the microbic flora in her intestine) she may transmit bacteria that are not wholly suitable for the child. It has been observed that the microbiota present in the milk of mothers who had undergone an urgent Caesarean section is more similar to that of women who had a vaginal delivery, while it differs from that of mothers who gave birth with a planned procedure: in this case, stress probably plays an important and not negative role as we have always been led to believe. These data do not represent merely a series of experimental observations far from the clinic, but can also orient us toward possible practical measures: supplementation of the pregnant woman with prebiotics and probiotics, repairing of the neonate's microbiota with the mother's vaginal flora and the use of probiotics and prebiotics at birth.

Once again, there is a window of opportunity at the time of birth. We can see this in studies on germ-free animals that are kept free of bacteria: besides our behavior, bacteria may also influence the structure of certain parts of the brain (e.g., the hippocampus), but only if they are present in a certain early period of life and not afterwards.

Metabolomics of mother's milk and formula milk

The more we study mother's milk, the more we have to admit our surprising lack of knowledge about this liquid, considered miraculous from ancient times. The metabolomic analysis of milk, which as a liquid is quite appropriate for this kind of examination, may contribute to expanding our knowledge of it. In a study on the metabolomics of milk using nuclear magnetic resonance and gas chromatography-mass spectrometry, we have observed that mother's milk presents high levels of lactose and low levels of maltose when compared to the formula milk commonly recommended and used in the neonatal period. The two types of milk (maternal and artificial) also present different levels of oleic and linolenic acid. Moreover, the milk of mothers of children born very early is different from that of mothers of late preterm children. Comparing the metabolomic data of mother's milk at 1 and 4 weeks, we see that numerous substances vary in quantity during the first month of life: oleic, linoleic, and palmitoleic acids increase as cholesterol decreases. Finally, new studies have made further important contributions that have led to the finding of certain milk characteristics due to genetics ("secretory" and "nonsecretory" mothers) or drug treatments.

Multipotent stem cells of mother's milk

We also know that mother's milk contains stem cells whose function is not yet fully known, but it is not difficult to imagine that they may trigger the beginning of the immune response of the newborn organism. In particular, it is thought that its colostrum is quite rich. The stem cells of mother's milk can be reprogrammed in the laboratory under the effect of special differentiating "cocktails" for many kinds of human tissues and may also be used in regenerative medicine. In reality, breastfeeding is itself a stem cell therapy. It may be that stem cells of mother's milk pass through the intestinal barrier, as well as that of the brain, and migrate to settle in the brain and other organs and tracts, among which certainly the thymus, pancreas, and liver. Once they reach the neonatal tissue of destination, they can differentiate to assimilate and integrate in such tissues. In the brain, stem cells can differentiate into neurons, astrocytes, and oligodendrocytes. We still have many mysteries and "miracles" to discover in mother's milk.

Mother's milk with garlic or curry?

An interesting and curious note: at birth, neonates can recognize olfactory and gustatory stimuli. The neonate immediately recognizes the mother's odor, which is a powerful stimulus to search for the mother's nipple, and this represents a factor of success in breastfeeding. Neonates usually dislike, or rather despise, the taste of garlic and curry; but if the mother during pregnancy eats garlic or curry regularly, they do not show this aversion. It is interesting to note that the orienting of taste from the culinary standpoint begins during life in the uterus. Within 2 minutes, the odor of colostrum increases the neonate's brain blood flow at the level of the olfactory lobes, whereas a bad smelling substance reduces it considerably. It is singular that vanilla produces the same effect as colostrum. Could this be where our weakness for sweets comes from?

Personalized fortification of milk of mothers who gave preterm birth

The growth rate of a preterm fed with the milk of its own mother is initially superimposable over that of the fetus in the womb. But after 2 weeks of life, with the progressive decrease in protein content, mother's milk appears to be insufficient for the nutritional requirements of a preterm weighing less than 1500 g at birth. In preterms fed only mother's milk we may see poor nourishment and scanty growth. For this reason, today's guidelines state that we must fortify the milk of these mothers. Fortification can be standard or personalized. Standard fortification does not take into account the basal characteristics of every single maternal milk and this means that an empiric amount of the different components is administered, and this does not always correspond to the requirements of the individual child. Personalized and customized fortification, following an analysis of the nutrients present in the milk, provides a preterm with milk of its own mother fortified on the basis of its actual nutritional needs. The advantages of personalized fortification consist of a qualitatively and quantitatively better growth of preterms even when they are of very low weight. Moreover, these neonates are hospitalized for a shorter time and this provides a cascade of benefits: it strengthens the mother–child relationship, which premature birth had interrupted too soon; mother and child are in the natural environment of their own home; the risk of contracting nosocomial infections is reduced and hospital costs are lower.

The fetus, the preterm baby, and the term baby: the good and the not so good

The "weight" of prematures

Every year in all the world approximately 15 million babies, about one out of ten, are born preterm: seven million in Asia, five million in Africa (with the highest percentage in the world, almost 12%), and just under five hundred thousand in Europe (6.2%). The four nations with the highest number of preterm births are India (3.5 million), China (about 1.2 million), followed by Nigeria and Pakistan (with approximately 750,000 each). Recently, there has been a global increase in preterm births, mostly caused by artificial insemination, multiple pregnancies, factors connected with maternal age, infections, nutritional issues, and postabortion complications.

In the last few years we have been accumulating a great deal of information as the result of basic research and clinical studies that show that intrauterine infections are closely related to fetal inflammation. Moreover, intrauterine infections are implicated in the early beginning of labor and play a central role in the onset of preterm neonate pathologies, such as bronchopulmonary dysplasia and damage to cerebral white matter. It is thus no surprise that metabolomic data can most likely predict and identify right from birth some of these pathologies before their diagnosis at a later date, which is the situation today. Finding the space-time coordinates of endouterine inflammation and its individual differences will contribute to strengthening our attempts to prevent preterm births and leaving the fetus in its proper environment for 9 months. The directions of research in the next few years will include the study and application of the -omics technologies as well as in-depth studies on the microbiota, maternal genital infections and inflammatory processes. Another element to focus on is the fine-tuning of epidemiological studies that compare data on mothers of different ethnic groups and economic conditions. Finally, it is important to increase the survival of preterms by keeping their short- and long-term negative outcomes to a minimum.

Survival depends almost entirely on the week of gestational age at birth: a US study published in *Pediatrics* in 2010 indicated that survival is about one in twenty at 22 weeks, one in four at 23 weeks, just over one in two at 24 weeks, and three in four at 25 weeks. We can say that starting from the 1960s the threshold of survival has been lowered by 1 week every 10 years. Inevitably, issues of a bioethical nature arise and one question prevails: what is the limit of viability, the aurora of life, the impalpable borderline between the spark of life and "nonlife"? Another issue of growing importance is represented by cost: it is estimated that saving the life of a preterm born before

Metabolomics and Microbiomics. http://dx.doi.org/10.1016/B978-0-12-805305-8.00007-8

the 32nd week and maintaining it until discharge costs an average of $280,811. But most of all, it is the asymmetry in the distribution of economic resources in the world's different geopolitical realities that impacts on the treatment of these babies.

The new inhabitants of the planet Earth: adults who were born with extremely low weight

In the last few years a new situation has arisen: today there are young adults who at birth weighed less than a thousand grams, which is known as extremely low birth weight (ELBW), and who survived. Up to not so long ago this was not the case, so they are the first of our planet's inhabitants born with extremely low weight who are walking the Earth after the approximately 10,000 generations of human beings that have preceded us. The concepts I have spoken about thus far on perinatal programming lead me to place the accent on the involvement of many organs in a situation that is objectively different from that of neonates born at term. I must therefore emphasize the need for a careful follow-up of these young adults so as to apply the best measures some optional, others mandatory, to safeguard them.

I do not want to discuss the issue at length here, but I merely wish to suggest some ideas to reflect on. I will speak of three organs of the premature neonate: the brain, the heart, and the kidney.

First the brain, since it is the most complex organ we know: when we begin to study it, it is like entering a never-ending tunnel from which we will not emerge. It is our most important organ and is acknowledged as such by our organism, which adopts all available strategies to protect it. In the case of fire, museums have a plan for evacuation with an order of priority of works to rescue; at Milan's Brera Academy let us say that the work considered the most precious is the renowned *Montefeltro Altar-piece* by Piero della Francesca, known more simply as the *Brera Altar-piece* (if I mention the ostrich egg hanging from the ceiling in it, it is easier to recall it) so it will be the first to be taken to safety. For our organism the brain is like the first work of art to be saved and put in a safe place.

Then I will speak of the heart and kidneys, both because they are in the top positions of the hit parade of important organs and because they always go together hand in hand (we speak of the "cardiorenal syndrome").

But before going into detail on programming of the different organs, I want to stress two concepts. The first is that we must be aware that the organs of a neonate, especially a preterm one, are completely different from those of an adult and a child. The differences are so great that from the histological standpoint they appear almost as if they were not the same organs. Paradoxically, there are disquieting similarities between the markers of fetal tissues and those of cancer cells. The second concept is that our therapies can modify the course of development of these organs. For example, if I treat a very early preterm with drugs that are necessary for its survival but at the same time are potentially nephrotoxic, the morphofunctional maturation of the kidney may change with respect to its normal development.

Cerebral programming

In recent years new evidence has been accumulating on the role played by the different factors that intervene in prenatal life to model the brain and cause in adulthood neurodegenerative diseases such as Parkinson's and Alzheimer's diseases.

The 9 months of growth in the womb and the first two or three years of a child's life are periods of the utmost importance and for this very reason are vulnerable and critical, not lastly owing to the strong increase in neuronal and glial cells and their connections.

Just think: one cell of the cerebellum can be connected to at least a thousand cells of the thalamus and one cell of the thalamus to at least a thousand cells of the cerebral cortex.

Many perinatal factors act epigenetically (a word with which we are now familiar), in particular, maternal diet, fetal malnutrition (under- or overnutrition), maternal stress, maternal hypertension and diabetes, prenatal infections, fetal hypoxia, prematurity and low birth weight, drugs administered to the mother, neurotoxic substances assumed by the mother (cocaine, amphetamines, cannabis), lack of certain trace elements (copper, zinc, iron), placental insufficiency, pollution, and chemicals in the environment (herbicides, pesticides, insecticides, dioxins, heavy metals, polychlorinated biphenyls, and organohalogen compounds).

Being born preterm interrupts maturation of the brain, which up to that time was proceeding in the right place at the right time.

As concerns degenerative diseases, the "two-strike" hypothesis states that a prenatal and/or perinatal insult to the brain may influence resistance (resiliency) or susceptibility (fragility) to the onset of Parkinson's or Alzheimer's disease many years afterwards. For example, it is not the same to have a third fewer or a third more neurons than average (once again we see the extraordinary variability between one person and another). So the practical corollary of this information is that prevention of neurodegenerative diseases begins in the womb. This message must be divulged and future mothers should be told to protect their child's brain starting from pregnancy.

Epigenetic factors that influence development of the brain: alcohol, aluminum, prematurity

Some of the epigenetic factors that impact on the brain's development are presented in Fig. 7.1. I will discuss three that are particularly widespread and unfortunately often underestimated: alcohol, aluminum, and prematurity.

Alcohol

The assumption of alcohol during pregnancy, whether in small amounts over a period of time or in large amounts from time to time (binge drinking) damages the fetus. At the same total amounts, high peaks are the most damaging compared to separate smaller quantities. Alcohol damage emerges at the brain level in the form of cognitive and behavioral disorders.

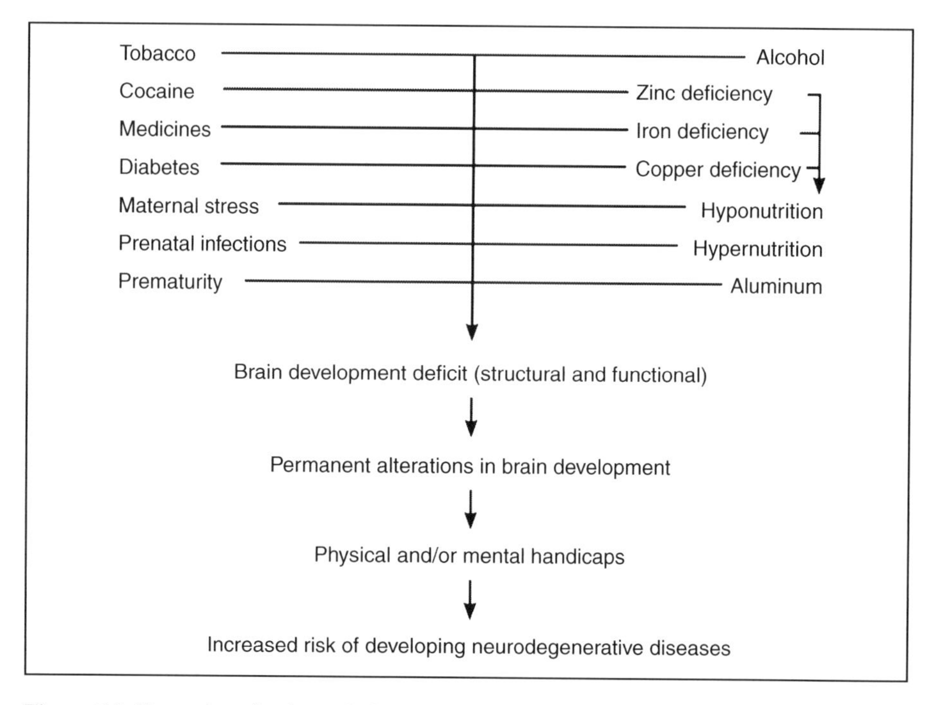

Figure 7.1 Examples of epigenetic factors that influence development of the brain. Modified from Piras et al. (2014).

Contrary to the past, today we know that even small amounts of alcohol can produce deleterious effects in the fetus of an expectant mother if she is slow to metabolize it. Alcohol consumption also goes hand in hand with other harmful habits such as smoking, the use of cannabis, and other narcotics. There is no alcohol threshold in pregnancy below which the fetus is not involved: for this reason pregnant women must avoid all alcohol consumption. Recently I visited the Gemäldegalerie in Berlin and found interesting a painting by Jan Steen entitled *The Christening*, dated 1663. In appearance, it portrays a pleasant family scene, with an unweaned baby sleeping in a wooden cradle in the foreground. Under the cradle is the proverb, "Youths chirp like old people sang." All seem to be enjoying themselves, but on closer look we can see many, too many, people drinking wine, both adults and children. Even the baby's mother is drinking, as she surely did when she was pregnant: how much has the just-christened child's brain suffered?

Aluminum

Embryo and fetus are at high risk of accumulation of aluminum, which is particularly harmful for the developing brain and bone structure; the same is true for the neonate, especially if preterm. The harmful effect may take place through interference in the amount of intracellular calcium present or through the capacity of aluminum to augment oxidative stress (caused by oxygen derivatives) and result in programmed cell death (apoptosis), in particular at the cerebral level. Aluminum can cross the blood–brain barrier and accumulate in glial cells and neurons. We know that the blood–brain barrier is all but a barrier and this is even truer the younger the child. In average adults there are approximately 600 km of blood vessels wrapped around nerve cells. The brain's blood vessels are coated with endothelial cells, but are apparently arranged more compactly than the vessels in other parts of the body. The microglia's task is to repair damage and its malfunctioning may correlate with long-term neurological problems. Preterm neonates are at risk of aluminum toxicity owing to immaturity of the intestine, the most important aluminum filter, and to immaturity of the kidney, the most important organ for excretion of this metal. Moreover, both preterm and term neonates who receive parenteral nutrition are at risk owing to the presence of aluminum in parenteral therapies (in any case necessary for their survival). Aluminum goes directly into the bloodstream and brain, bypassing the intestinal barriers. Finally, babies with kidney insufficiency are also at risk of aluminum toxicity, owing both to reduced excretion of the metal and the concomitant use of antacids. Since little is said about this subject, here are some practical preventive suggestions for mothers during pregnancy and while breastfeeding:

- avoid acid beverages or tea packed in aluminum cans; in particular, avoid fizzy drinks that "bite" the can's aluminum and dissolve it in the beverages;
- limit the consumption of tea and use milk instead of lemon (lemons increase aluminum absorption whereas milk reduces its bioavailability);
- limit the use of toothpastes and antiperspirants that contain aluminum;
- avoid coffee made in aluminum coffee makers;
- limit the consumption of foods cooked or preserved in aluminum containers;
- avoid cheeses or sweets containing aluminum (some food additives contain aluminum);
- avoid the use of medicinal herbs for which the level of aluminum contamination is not specified;
- avoid the use of cosmetics containing aluminum;
- dilute formula milk with mineral water that has the lowest aluminum content;
- choose formula milk with the lowest aluminum content;
- limit the use of soy formulas owing to their high aluminum content;
- limit parenteral therapies owing to their high aluminum content;
- avoid medicines containing aluminum, especially antacids;
- do not neglect or postpone vaccination with vaccines containing aluminum since their content of this metal is low.

Prematurity

A quite special problem is the long-term neurological outcome of babies born preterm with very low weight; I will mention this briefly, since it is much better known than those I discussed previously. A severe disability at 18–24 months has been observed in one out of three ELBW neonates weighing less than 1000 g, with multiple disabilities in one child out of four. Thus, the follow-up of these children is absolutely obligatory. Other disorders have been described: the autism spectrum disorders, the attention deficit hyperactivity disorder (ADHD), hyperactivity, emotional and mood disorders, in particular depression. Finally, other problems, such as elusive unspecified neurocognitive disabilities, have been observed in 50–70% of ELBWs without disabilities and of normal intelligence. They often emerge after the beginning of school activities when they are asked to perform.

Cardiac programming

It has been known for a long time that we can find a cardiovascular pathology even in young people. As far back as 1915, the German pathologist Mönckeberg wrote that the autopsy reports on 140 soldiers killed during the First World War presented atherosclerotic plaques in their coronary arteries in 46% of cases. Enos and coworkers performed autopsies on 300 young soldiers killed in the Korean war: about 80% of them presented atherosclerotic elements in the coronaries and 3% presented occlusions in more than 50% of the coronaries.

There is a gender difference in the formation of atherosclerotic plaques. Women in the fertile age are protected by estrogens, which have a vasodilator effect that widens and elasticizes the vessels; atherosclerotic plaques in women usually form after menopause in relation to the decrease in hormones. Everything most probably starts from the first moments in the womb. From the 16th–35th week of pregnancy the linear increase in heart volume is caused entirely by the proliferation of cardiac cells, the cardiomyocytes. The net production of cardiomyocyte nuclei is above two million per hour, numbers that make the head spin! At the end of pregnancy and in the early neonatal period there is a change from hyperplastic cell growth (many cells) to hypertrophic cell growth (larger cells).

Many are the pre- and perinatal factors that can impact negatively on the processes mentioned above: fetal hypoxia, prematurity, intrauterine growth restriction, undernutrition, maternal autoimmune disease such as systemic lupus erythematosus, substance abuse, and maternal smoking. These factors remodel the heart. In particular, we know what takes place in babies with restricted intrauterine growth. If I were to describe this in few words, I would unfortunately have to choose "less efficient hearts." These children's hearts tend to dilate, have a less efficient contractile capacity and a vascular dysfunction associated with an increase in the thickness of the carotid artery wall, and high blood pressure. Children with restricted intrauterine growth run a greater risk of developing long-term hypertension and kidney damage than those born preterm. Among preterm neonates, those who are also of low weight for their gestational age are at higher risk of metabolic disorder.

In Table 7.1 I present the problems in apparently healthy young adults who were of extremely low weight at birth. These youths tend to suffer from arrhythmia, their blood vessels are less elastic and extensible, which is to say rather rigid, inside of which the substances that induce stricture prevail. All of this favors the onset of high blood pressure. Besides the suggestions for treatment shown in the table, in infancy and adolescence an appropriate lifestyle is to be recommended (physical activity, avoidance of passive smoking, and avoiding obesity). In any case, we can never say enough about the need to intervene preventively and recommend a careful prenatal diet for expectant mothers.

Table 7.1 Possible long-term consequences in adulthood to subjects born with extremely low weight: suggestions for diagnosis and care

Possible consequences in adulthood	Risks	Suggestions for diagnosis and care
Increase in the QT interval of the ECG in some subjects	Risk of arrhythmia and sudden death	ECG monitoring Avoidance of drugs that increase the QT interval
Reduced vascular elasticity	Risk of hypertension	Blood pressure monitoring
High ADMA levels	Risk of future acute cardiovascular problems	Ecocardiography and blood pressure monitoring
Increase in microalbuminuria and urinary NGAL, reduction of kidney volume	Risk of chronic kidney insufficiency	Urine stick monitoring, albuminuria, creatinine and cystatin C in the blood, kidney ultrasound

QT: an electrocardiogram interval which, if it increases in time compared to normal, may cause arrhythmia and sudden death; ECG: electrocardiogram; ADMA: asymmetric dimethyl arginine, a powerful inhibitor of nitric oxide with vasoconstrictor effects; NGAL: neutrophil gelatinase associated lipocalin, a marker of kidney injury.
From Bassareo et al. (2013).

Kidney programming

Can glomerulogenesis (the formation of glomerules), which normally comes to an end at 35 or 36 weeks of gestation, continue after birth in an extremely premature neonate? Can the unnatural environment in which a baby with these characteristics finds itself alter glomerulogenesis and cause a reduced final number of nephrons? How is the kidney function of these babies a long time after birth?

Preterm birth, especially if extremely preterm, can by itself negatively influence nephrogenesis and glomerulogenesis. It is not known to all that nephrogenesis continues for a maximum of 6 weeks after a preterm birth. So if a baby is delivered prematurely at 24 weeks of gestational age, its kidney may have an increase in the number of glomerules for 6 weeks only: at 30 weeks of postconceptional age (twenty-four plus six after birth), glomerulogenesis for that neonate stops and cannot continue to the 35th or 36th week, the natural term for kidney development in a regular pregnancy. Similarly, if a baby is born at the 26th week of pregnancy, its kidney may increase in the number of glomerules only for another 6 weeks from birth and glomerulogenesis terminates at 32 weeks of postconceptional age. If the neonate has suffered an insult, such as asphyxia, that has caused acute kidney injury, this period of kidney maturation is shortened. The same can be said if the neonate has had an intrauterine growth restriction. In the neonatal period no problem may come to the fore, but when the child grows, the gap widens between the work of removing wastes to be done and the quantity and quality of the work actually performed by the nephrons, with the consequent onset of chronic kidney insufficiency. Imagine a town of 10,000 residents with a waste disposal system that uses 20 trucks. If the town grows and reaches a population of a hundred thousand, the same number of trucks will no longer be able to collect all the trash.

Neonatal ethics as a paradigm

At this point, considering everything that may occur to highly preterm neonates and those of extremely low weight, some may wonder if treating these children is sensible and if it is ethical to do so.

In May 1974, a pioneering event took place: at a ranch in the Valley of the Moon in California, a meeting took place with some 20 experts in different disciplines (neonatologists first of all, but also experts in law, theology, philosophy, and social sciences) for the purpose of discussing two basic questions: is there a right not to resuscitate a neonate at birth? and, is there the right to discontinue life support after a clear and univocal diagnostic assessment? The discussion was on two fundamental points: the withholding and withdrawing of treatment. We are always in precarious equilibrium: we must avoid causing injury by not intervening, but we must also avoid intervening when it is impossible to support the life and health of patients without excessively increasing or prolonging their agony. Leaving aside the question of faith, I believe it is interesting to quote a passage from the *Declaration on Euthanasia* of the Holy Congregation for the Doctrine of the Faith of May 5, 1980, taken from John Paul II in his Encyclical *Evangelium Vitae* of March 25, 1995: "When inevitable death is imminent in spite of the means used, it is permitted in conscience to take the decision to refuse forms of treatment that would only secure a precarious and burdensome prolongation of life, so long as the normal care due to the sick person in similar cases is not interrupted. In such circumstances the doctor has no reason to reproach himself with failing to help the person in danger".

Some years ago this subject was discussed at length concerning highly preterm neonates at the inaugural round table of the XIII Italian National Neonatology Congress in Rimini. To best understand the climate, here are some of the comments made during the discussion: "Are we doing experimental medicine?"; "Neither the triumph of hope over reason, nor the victory of the ego over uncertainty"; "Beware of the omnipotence of the neonatologist!"

The question that is asked in the literature today is: how small is too small? A baby who weighs 500 g at birth weighs one-sixth of normal weight: here we are at the frontier of life. But at what price victory, and what kind of victory? There are different ways to approach such issues: should we privilege the sanctity or the quality of life?

As you can see, this paragraph is full of questions, more questions than answers. But we can say that every human being has a right to the best possible quality of life he or she can reach. It is true that medicine is continually challenged in trying to define the limit of gestational age below which to withhold intensive care. On the other hand, the ontological statute of the preterm patient, even at the threshold of life, is that of an adult. There can be no strategy of negligence or inattention toward those principles of medical ethics that are recognized for adults.

The widespread application of new technologies, such as the -omics disciplines, while representing a fundamental step forward toward personalized medicine, may also represent the other side of the coin that brings questions to the fore: this mass of knowledge will increase the power of one person over another and, paradoxically, will raise crucial issues concerning the freedom that each person will enjoy in the future. The following sentence, translated from the entry "Progresso" of the *Grande Dizionario Enciclopedico UTET*, which dates back some years, is food for thought: "Today the quality of life is higher than in any past epoch, but perhaps this is also true compared to future epochs… We are living in an intermediate stage between amazement at the progress made and fear of its possible consequences."

In the next two paragraphs I will introduce some of the experiences that characterize perinatal life, because the mysterious life of the fetus and neonate must be understood and defended.

The wonderful and active world of the fetus

The fetus lives immersed in the amniotic liquid, it swims and floats in a warm and comfortable environment. It has tactile sensations as demonstrated by the fact that at 14 weeks of gestational age it reacts to pressures. Sensitivity to temperature and pain develop later. As concerns sight, the fetus perceives differences in the intensity of light, just as we do when looking through a thick curtain. The sense of smell also develops early, although it is not used much in intrauterine life; there is, however, olfactory programming with a long-term memory: after delivery, the neonate is attracted by the smell of colostrum and tries to reach its mother's nipple. The maximum stimulation of the olfactory lobe takes place when the baby receives colostrum. One of the proofs of the importance of smell, which it has always had in the evolution of the species, is the fact that a twentieth of human genes is associated with odors and that humans are able to distinguish more than 10,000 different kinds of smells. As Süskind states: "The persuasive power of an odor cannot be fended off, it enters into us like breath into our lungs, it fills us up, imbues us totally. There is no remedy for it."

The fetus has the sense of taste. We can say that fetuses like "candies": it has been observed that after an intraamniotic injection of saccharine the fetus drinks more amniotic liquid, thus causing its decrease. The opposite occurs when a bitter substance is injected.

Finally, we can mention one of the many reasons why the fetus is not a miniature adult. When adults are exposed to a danger, they react with "fight or flee" behavior: the heart beats fast, the breathing is that of a wild animal being pursued, the pupils dilate (midriasis), and the hair stands on end. All this is connected to the action of adrenalin and other stress hormones. These same hormones increase in fetal stress, but produce a response opposite to that of the adult: its heart beats more slowly and it stops moving, almost as if not to be heard: it "plays dead." It is how it tries to save oxygen and energy. We also speak of an "immersion reflex," like that of seals and ducks when they dive underwater.

The extraordinary world of sound of fetus and neonate

Auditory experiences in the perinatal period deserve a paragraph of their own. Although a series of concentric barriers (amniotic liquid, embryonic membranes, the mother's uterus and abdomen) separate the fetus from the outside world, the fetus lives in a world rich in sounds, vibrations, and movements. Voices and music reach the fetus without significant distortions. How does the fetus react to sounds? Fetal motor responses to acoustic stimulation have been documented with ultrasounds at 24 weeks of gestational age and are significant at 28 weeks, depending on maturation of the central nervous system. Perhaps not everyone knows that there are studies that report an increased risk of preterm delivery (one and a half times that of controls) and low neonatal birth weight if the mother lives near an airport.

The mother's voice, which is a magnificent bridge for communication between mother and fetus, is particularly powerful because the fetus hears it through her body. The fetus hears the mother's heartbeat from the inside, where sounds are prevalently at low frequency. Following birth, the neonate is pacified by being placed in contact with the mother, especially on her belly, where it hears her heart beating once again, this time from the outside.

The acoustic competence and preferences of a fetus are presented in Table 7.2.

We know that Vivaldi and Mozart calm the fetus while Brahms and Beethoven excite it. Rock and Roll music unleashes the fetus: at concerts pregnant women are sometimes forced to leave owing to the hard kicks of their unborn child.

Table 7.2 Favorite sounds of a fetus

- Perception of voices and music
- Finer perception of the mother's voice compared to other female voices
- Finer perception of the mother singing compared to her talking
- Finer perception of singing voices (or recorded opera arias) compared to voices that are not singing

Table 7.3 Beneficial effects of music on preterm neonates and their parents

Beneficial effects of music on the preterm neonate (from week 28 on):

- It improves oxygen saturation
- It favors the gaining of weight
- It increases tolerance of stimuli
- It reduces hospital time

Music has beneficial effects on parents:

- It increases the number of visits to the neonatal intensive care unit
- It favors the bond with the child
- It favors parental empowerment

Modified from Standley (2002).

Even after birth the child is not indifferent to music. Based on an experiment by Hugo Lagercrantz, the neonate who hears a piece by Mozart feels wonderful and breathes regularly. But if it hears Stravinsky's *Firebird*, it begins to have apnea crises.

A very recent article reports the pain-relieving effect of Mozart's music during the taking of routine samples from babies with low weight at birth.

The newborn child prefers the mother's normal adult voice to that of "baby talk": this is because of prenatal familiarity with it; it also prefers the idiom spoken by the mother during pregnancy over another language, even when spoken by the same mother, once again because of what it was used to before coming into the world.

The premature baby in intensive care receives stimuli that it did not receive in the uterus: these are visual and auditory stimuli at high frequency, much more intense than those it was used to. It also suffers from a deprivation of other stimuli such as the filtered voice of its mother. There is a substantial difference between silence, sound, noise, and music. We know quite well the negative effects of noise on the preterm neonate: sleep disorders, auditory disorders, somatic effects, effects on development of the emotions. On the contrary, music is beneficial to these babies and their parents, as can be seen in Table 7.3.

Plato, in his *Republic*, stated that musical education is more powerful than any other, since rhythm and harmony found their way deep into the soul, taking possession of it and imparting to those who had received it wisdom and reason.

We are an ecosystem: in our bodies only one cell out of ten is human

Brain versus gut: an ongoing war from the Pleistocene to the present

We do not live on what we eat, but on what we digest, as the saying goes.

In the course of human evolution, brain and gut have always been in competition. The brain has always come out on top (it could not be otherwise) and the history of evolution is there to prove it. But the gut may win the rematch and it has learned to influence our brain, if not to control it at least partially, through the microbiota.

A first demonstration of the "defeat" of the gut is represented by the evident modifications in our rib cage that occurred in the passage from *Australopithecus* to *Homo erectus*. These changes were made possible by a reduction of the intestine which made possible the passage from a rib cage shaped like an upside-down funnel to one shaped as in modern humans.

A second, even more important proof is that of the progressive encephalization of *Homo sapiens* and the asymmetric development of the weight of human organs. Adult humans devote 20–25% of their metabolism to cerebral activity (even 90% in the neonatal period and 60% in infancy), compared to 8–10% in primates and 3–5% in mammals other than primates. Thus, the brain of an adult human (about one-fiftieth of body weight) receives at least one-fifth of the energy. In human evolution, the increase in brain size and the consequent energy sacrifice in its favor have come about at the expense of the gut. If the weight of the liver, kidney, and heart has been more or less constant over the millennia, the brain's increase in weight, anatomy, and functionality has taken place at the expense of the intestine. These changes are the result of an improvement in diet, with an intake of high-energy foods of animal origin made more digestible by cooking: the lower amount of energy required for digestion has led to a reduction in the size of the intestinal tract and left more energy available for expansion of the brain.

The growth of the container (the cranium) has not always been associated with the growth of the content (the encephalon) throughout our evolution. Only in *Homo sapiens* has this process taken place, and for this to happen the intestine had to be sacrificed. At the same approximate weight of the liver, intestine, and brain, the latter burns three and a half times the energy as the other two organs. At birth, a term neonate has a brain that weighs about 300–400 g. In the first 1000 days of life the brain gains 1 g of weight per day on average; so, at the end of the first 3 years the brain almost reaches the weight of an adult brain. As concerns increase in the brain's mass,

Metabolomics and Microbiomics. http://dx.doi.org/10.1016/B978-0-12-805305-8.00008-X

what is supposed to take place has already taken place in the first 3 years of life. In this period, which may be one of vulnerability but also of opportunity, an optimal nutrition plays a decisive role.

Although in the past it has been a controversial matter, we now know that the encephalization of the adult is proportional to that of the neonate. Variations in brain size in mammalians correlate with life histories. The species with the most developed brains have longer pregnancies, they mature more slowly, and live longer lives. These situations have been explained not only in terms of the cost of development, since brains of larger size take longer to grow, but also of cognitive benefits, since larger brains increase the chances of survival of the species and the length of life. In support of the hypothesis of the cost of development, it has been shown that evolutionary changes in the growth of the brain correlate specifically with the duration of the relevant phases of the so-called "maternal investment," that is, gestation and breastfeeding.

The brain is a jungle, not a computer

At the end of growth, the adult human brain consists of about 100 billion (10^{11}) specialized neurons: each neuron has from 1000 to 10,000 synapses for a total of about 10^{14}–10^{15} connections. Not all the many synapses of each cortical neuron are present at birth; on the contrary, they proliferate discontinuously in humans, in wave after wave from birth to puberty during the encephalization phase. An extremely thick and inextricable jungle, not a computer: this is our brain.

Besides encephalization, other elements that have contributed to the evolution of the human brain are represented by its hierarchical architecture and plasticity, which can be seen through connectomics.

The connections of the different parts of the brain are organized, even in the neonate, as hubs and spokes, which is to say scale-free network systems. The axons and dendrites (the ramifications) of the neurons (the nerve cells) increase after birth because the synapses form in relation to the experiences a person has had. We know that in the neonate and unweaned baby about one million synapses form every second in the first months of life: learning means consolidating preestablished synaptic combinations and eliminating surpluses. What takes place is a transitory redundancy that derives from an initial exuberant growth of neurons and synapses followed by a synaptic pruning, with the cutting off of some of the synaptic branches. For this reason, as adults we have about 60% fewer synapses than we had at 2 years of age. Only those that are actually exploited by the organism are retained following the rule of "use it or lose it." If to reach the seashore we can follow three different paths through the dunes. The first one is striking: it passes alongside a marsh with flamingoes. The second is also attractive, but a bit less than the first. The third is less beautiful: it is rocky and there are cottages that spoil the view. Every time we go to the beach there, we choose the first path and sometimes the second, but never the third. The first one will always be used and will be well-kept and beaten, the second a little less, and the third will be buried in weeds and sand because nobody uses it. What is it that decides which synapses to keep and which to discard? The answer is the environment, mother's milk, mother's cuddling. More completely, we can answer in this way: the satisfying of our five basic needs (eating, drinking, sleeping, communicating and, when older, sex). Thus, the environment and stimuli model the brain after birth. The synapses are specialized and are kept when they are stimulated repeatedly. The first experiences are quite important and the moments of attachment are fundamental for brain development. Attachment is the primary architect of the brain. Emotional development, on which depends our cognitive progress, takes place in the first 2 years of life. So what we thought in the past, that our brain was correlated only with our genes, is false: today we know that it depends on the interaction between genes and environment and our first experiences in life. These experiences convert into choices and the maintaining of cerebral connections. Another legend that has been disproved is that the brain of the adult works harder than that of a child. The opposite is true: the brain of an unweaned child works twice as hard as that of an adult, and not only in waking hours. Even during sleep the synapses form. To sleep and sleep well is thus quite important, especially for children. There is even a popular saying to this effect: "The baby that sleeps grows."

Prematurity and asphyxia are among the causes that may bring about a loss of neurological function by significantly reducing the connections of the different parts of the brain. Neurological injury can also lead to neuronal death through a mechanism that activates the microglia. Many neurological pathologies are alterations of our connectome, which is an extraordinary example of ductility since it changes in continuation throughout our lives in accordance with the four following modalities: reweighting, reconnection, rewiring, and regeneration. The first is reweighting by the neurons of the connections that can be strengthened or weakened; the second is reconnection, which occurs by creating or eliminating synapses; the third is rewiring of the neurons that can make the neuronal ramifications grow or shrink; the fourth is the elimination or creation of new neurons. This incessant activity should require enormous amounts of energy, but what shocked was that the neuroscientist Michel Hofman, cited in Swaab's bestseller, calculated that the cost of all the energy consumed by the brain in the life of an 80-year-old amounts to about $1500 at today's prices. What computer could do all this at this price for such a long time?

The brain–gut connection and microbiota: the gut's revenge?

Let us start from the title of an article that appeared in *The Economist*: *The human microbiome: me, myself, us*. An increasingly large group of biologists see human beings not as single individuals but as ecosystems. Only one cell out of ten in our bodies is a human cell. The human body is a superorganism in which thousands of microbic genomes interact continuously among themselves (*sociomicrobiology* is spoken of) and with the human genome. This statement has to make us think deeply. We have an ecosystem, a world inside of us of which we are almost entirely unaware. Who controls it? To what extent are we influenced by this unknown population living inside us and outside of us, which has an incredible complexity and interconnectivity? Considering people as ecosystems containing many species, some of which supportive and others antagonistic, could even lead to a change in medical practice and more in general in the medicine of the future. Indeed, the characterization of the human microbiota in different disease conditions suggests that our microbic environment plays a crucial role in maintaining health and in the onset of many diseases, including those of the nervous system. Many inflammatory diseases seem to make their appearance in relation to a brusque change in microbiota composition or to upsetting of the delicate balance among the thousands of microbic communities. Some of the diseases characterized by a dysbiosis are Crohn's disease, ulcerative colitis, the irritable bowel syndrome, psoriasis, diabetes, and cardiovascular diseases. In practice, inflammatory diseases are not caused by contact with a single pathogen: Koch's postulate (one microbe—one disease) is no longer necessarily true in the era of metagenomics or post-genomics, terms that include everything that comes sequentially and chronologically after genomics. Another extremely important practical corollary is the following: the failure of many present-day therapies may be linked to the fact that we now treat the single individual as a separate organism and not as an entire ecosystem out of balance. The medicine of the future will be forced more and more, willingly or unwillingly, to come to grips with this concept.

In reality, human beings are born and develop in a microbic world as *Homo bacteriens*. We could even say that *Homo sapiens* is a support or vehicle for other dominant life forms, albeit small in size. The overall genome of bacteria is composed of over 3 million genes, at least 100 times more numerous than our human genes.

In many ways, the intestine is still an unknown organ, one that is little exploited compared to other organs, even though it is called "the second brain." To emphasize its importance, some authors (a bit provocatively) state: "In gut we trust." The following numbers emphasize its key role: in the gut there are 100 trillion bacteria (ten times the number of cells in our bodies) belonging to 1000 different species; considering only the small intestine, there are more than 100 million neurons; 70% of the immune cells of the organism live there; the absorbing surface totals 300 m^2. The presence of an immense number of neurons (their largest "repository" after the brain) bears witness to the gut's close connection with the brain. We know this even at the level of popular wisdom. If we receive bad news, or we are tense or excited, our intestine is immediately affected. A very recent article has in its title the expression *gut emotions*.

The control room of our organism: the neuro-endo-immune supersystem

Some authors believe that the control room of our organism is represented by the neuro-endo-immune supersystem. An issue of extreme and growing interest is the impact of the microbiota on the immune system. The microbiota is a true extra organ which in adults weighs about 1500 g (more than the brain). Its function is protective, metabolic, trophic, and immunological. In particular, by controlling the immune system the intestinal microbiota (also formally called the intestinal microflora) is in some way capable of controlling our brain. Maturation of the intestine and its microbic colonization take place following birth in parallel with other crucial events such as development of the immune system and cognition. Moreover, the bacteria appear to be involved in energy accumulation and blood pressure regulation. Finally, the creation of an ecosystem in an early period of life plays a key role in significantly orienting proneness to disease throughout an entire lifetime. Development of the intestinal microbiota is characterized by rapid and large changes in the quantity, diversity, and composition of microbes. These changes are influenced by medical, cultural, and environmental factors such as genetics (which we must not forget), the kind of delivery, diet, family environment, past diseases, and the treatments used in the single individual. It is thus impossible to establish a universal standard for intestinal colonization. The factors that impact on the development of a "tailored" microbiota in the neonate are well known: maternal chorioamnionitis, gestational age, kind of delivery (spontaneous or Caesarean), breastfeeding or artificial milk, neonatal diseases, early antibiotic treatments, and presence of siblings in the family. A qualitative and quantitative disorder in intestinal colonization (dysbiosis) is associated with alterations in the immune response. As concerns the perinatal period, we know of interesting developments in knowledge concerning the role of the maternal microbiota in determining preterm delivery and delivery more in general. I can mention the influence of periodontitis or urinary infections on preterm deliveries. In particular, many studies that correlate preterm delivery with periodontal diseases have been published. There is a great deal of literature on the role of periodontal diseases in causing negative results in pregnancy, such as preeclampsia, prematurity, and low birth weight. The mechanisms that have been identified are both direct—the microorganisms of the oral cavity reach the fetus-placental unit—and indirect, through the cascade of mediators of the inflammation. This fact coincides perfectly with the practically exclusive concordance between the placental microbiota and the bacteria of the oral cavity, tongue, throat, and pharynx, as I mentioned earlier.

An essential issue is to establish at what age the child's microbiota becomes similar to that of the adult. Recent data on large cohorts of babies in different parts of the world show that this occurs at about 3 years of age. To say the least, it is surprising that this is the same age at which the human brain is consolidated.

Studies are comparing the microbiomes of different individuals in different normal and pathological situations. Here is a brief outline of present-day knowledge and some future developments.

Three different enterotypes

The intestinal microbiota can be defined as a consortium of groups of bacteria. This consortium coexists with the host from the time of birth and grows with it to form an authentic, advanced superorganism, interactively modulated and in true symbiosis with the several components. Here, as said before, we are dealing with sociomicrobiology. This information can be useful in better understanding how to fight disease-bearing and multiresistant bacteria. It has been shown that there are three different enterotypes in the human intestinal microbiota. At present, we do not know if, and which, genetic and environmental factors select them. Furthermore, the fecal samples by means of which their characterization has been performed are not necessarily representative of the entire intestinal ecosystem. Finally, the technology of the future will enable us to define the individual enterotype correlated with the state of health and illness of a single person and will have a bearing on different responses to diet and drugs. On the basis of these acquisitions we will be able to identify functional markers of individual responses that will be extremely useful in many conditions, such as metabolic syndrome and colon-rectal cancer.

How to change and monitor our enterotype

We can imagine, as we did in the case of metabolomics, that we will soon have in our homes an economical kit for monitoring the microbiota, for example, during a relapse of an infection of the urinary tract, or during antibiotic or antitumor therapies: we may be able to gain useful information in real time on what to do for the single patient. Today instead we can find Western children who arrive at the age of 18 after going through no fewer than 15 cycles of antibiotic treatments (not always beneficial), each of which alters the intestinal flora even to a significant extent. In the future, we will be able to personalize our diets in accordance with our intestinal bacteria (our personal enterotype), or even perform a microbiota transplant from a healthy person to one who is ill, thus modifying the guest bacteria in favor of the latter. But even now we can change our eating habits: we can eat large amounts of vegetables as grandmothers in the past recommended. We know this suggestion is useful because when our "good" intestinal microflora digests and metabolizes vegetables, we have the formation of butyrate. This substance has properties that attack infections and act epigenetically on our genes. By manipulating our diets we can modify our intestinal bacteria. In the adult, this process takes longer than it does in children and neonates. We can also use probiotics ("bacteria for food"), prebiotics ("food for bacteria"), and symbiotics (probiotics plus prebiotics). A treatment that will certainly develop in the future, although now it is only at the beginning, is the transplant of intestinal microbiota (or fecal microbiota transplant—FMT) from a healthy person to a patient. In particular, the scientific community is acknowledging that the microbiota is functional to the normal physiology of the organism and that FMT represents a new kind of therapy. Feces, after proper processing, can be transferred by probe via nose or rectum: the new bacteria take root quickly and replace a part of the old ones. We can say, with no fear of being contradicted, that at first the procedure appears decidedly disgusting. But the results are promising, if not spectacular in cases of adults with serious and recurrent colitis resistant to powerful antibiotics such as vancomycin, or in chronic intestinal diseases such as Crohn's disease and ulcerative colitis. We can even imagine in the case of obese persons the removal of gut bacteria that take large amounts of energy and calories from food and their replacement with others having the opposite characteristics from thin people. Still more important is the correlation between intestinal microflora and tumors, especially that of the colon.

There is a close affinity between microbiota and the fecal metabolome. I mention here necrotizing enterocolitis, an extremely serious neonatal intestinal pathology that may require surgical procedures that are devastating for the intestine and may lead to death of the neonate. Deaths resulting from this disease have increased in recent years despite the progressive decrease in mortality from other causes such as lung diseases, immaturity, and brain infections and injuries, as the result of a study published in the prestigious *New England Journal of Medicine*. The expression "necrotizing enterocolitis" means an inflammatory and infective intestinal pathology that mostly involves the colon. It leads to necrosis (death) of the intestine. One child out of three requires surgery and one out of ten dies. One of two cases occurs even when the neonate is fed mother's milk only. At the onset of necrotizing enterocolitis there is an abnormal colonization by *Clostridia* spp., which may lead to a modification of metabolites.

We are now entering the next part of the chapter: microbiomics and metabolomics, two magic words. Concerning microbiota, prebiotics, and probiotics, there is today a gaping deficit in communication, but even more in knowledge, among those who practice medicine "in the front lines." This lack is even more evident in the emerging and expanding world of the -omics technologies. Microbiomics has to be properly explained to patients, physicians, nurses, pharmacists, politicians, and other stakeholders in the field of health care.

Microbiomics and metabolomics

A specific area of growing interest, one that is being addressed more and more frequently, is the correlation between microbiota and metabolomics. In particular, this is the application of metabolomics in deciphering the metabolic impact of the intestinal microflora on health and illness. We are certain that there are special relationships between microbiota and metabolome and that metabolomic analysis reveals noteworthy effects of the intestinal microflora on the blood metabolites of mammals. From the study of germ-free animals, it clearly emerges that the microflora has a direct impact on the host's capacity to metabolize drugs, and this suggests a close reciprocal action between bacteria and the mammalian metabolism. In subjects genetically prone to a disease, only with a certain diet and a specific microbiota can we hope to avoid the formation of specific metabolites capable of damaging the organs of an organism, including the brain (the so-called brain–gut connection). Bacteria are many, they are extremely small, they preside over all areas of our organism, they are organized in villages (Fig. 8.1), they are practically invisible, and yet they are often in the driver's seat. When we treat patients, we should treat the entire bacterial ecosystem that lives in and around them, otherwise the treatment may be simply of the cosmetic kind. It is as if we kept on removing toxic fumes from inside and around a house without searching for their source and discovering why they are produced.

Where are the villages of the bacteria in our bodies? We can proceed by populations of bacterial species (in round numbers to help the memory): the large intestine has 34,000 species; the tongue has 8000; the throat has 4000; nostrils, the retroauricular

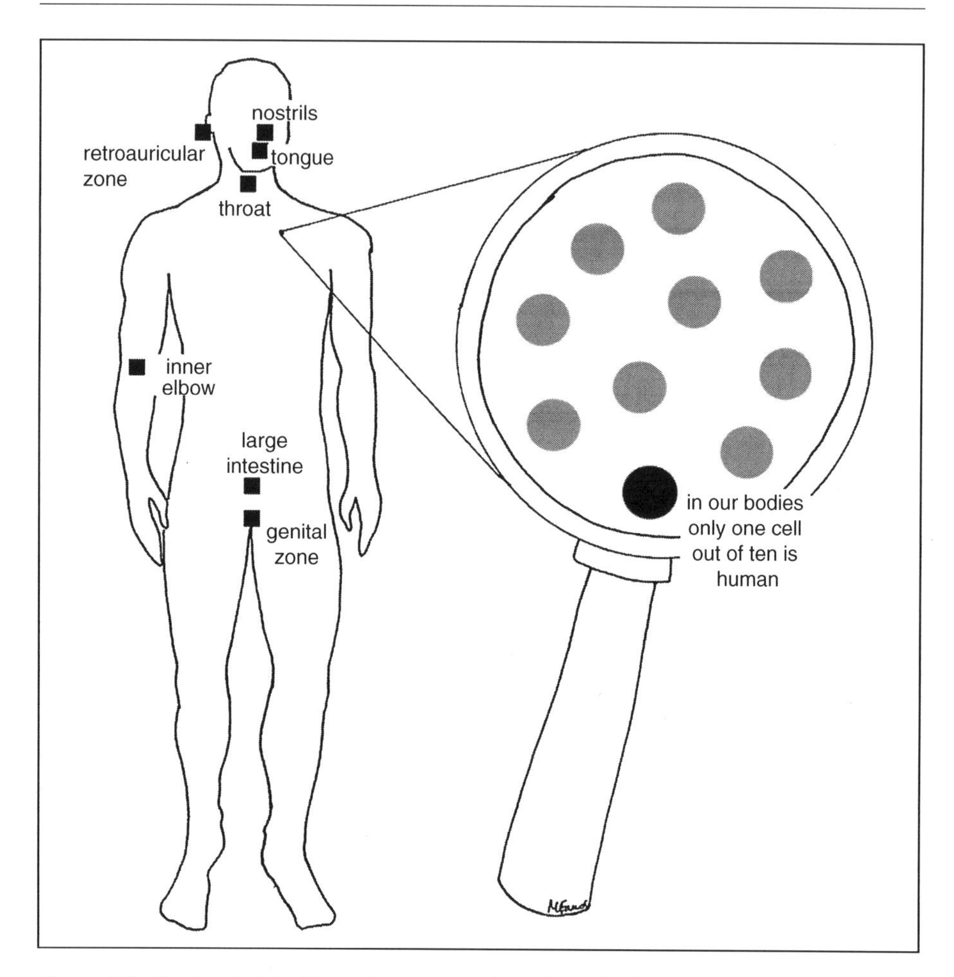

Figure 8.1 The bacterial villages in our organism. In our bodies only one cell out of ten is human.
Modified by Margherita Fanos from Wolfe (2013).

zone (behind the ears), inner elbows, and genital area 2000 species each. Some of these bacteria are dangerous inhabitants, some are authentic serial killers, but most are essential for us and without them we could not survive. They help us digest food and absorb nutrients, they protect the intestinal walls, they train, orient and guide our immune system, and defend us from the "criminal" bacteria, forming a sort of picket line that stops the bad bacteria from taking over.

Each village has a tribe that prevails over the others. In the vaginal aperture prevails *Lactobacillus acidophilus*, the producer of lactic acid: this keeps the local pH low and inhibits the growth of dangerous bacteria.

In the last decade, researchers have demonstrated a bidirectional relationship between the intestinal microbiota and brain functions: the gut–brain or brain–gut axis.

We know that they communicate in both directions by means of neural, endocrine, and immune mechanisms, and that the microbiota is involved in the response to stress and the development of the central nervous system during some critical windows of development. We can recall the slogan "food and mood." There is a link between eating well and wellbeing, mood, and anxiety. A practical corollary is: by improving nutrition we can improve treatment of cognitive and mood disorders. Some tryptophan derivatives, such as L-kynurenin, kynurenic acid, and quinolinic acid, play a leading role in this sphere. Therefore, modulation of the intestinal microbiota may become a new therapeutic objective in the prevention of mood and anxiety disorders. A practical example of how the microbiota strongly influences the brain is represented by the fact that orally administered nonabsorbable antibiotics inhibit the bacteria of the intestinal microflora and can cause a regression of encephalopathy in patients with a hepatic decompensation. Some improvements in behavior, albeit transitory and limited to the period of treatment, have been observed in autistic children treated with vancomycin.

Microbiomics: from the clinic… to the chocolate

From kefir to fecal transplant

The idea of the intestinal microbiota as the key regulator of the states of health and illness comes from the studies by Élie Metchnikoff performed over 100 years ago. He advanced the hypothesis that certain toxins produced by putrefactive bacteria in the colon accelerated senescence and that beneficial bacteria could replace the harmful ones. Metchnikoff noted that among certain rural populations of East Europe the consumption of kefir, a beverage of fermented milk, was correlated with greater longevity, so he added this to his diet, thus formally launching probiotic therapy.

Starting from these origins and after a long trek, today we have finally arrived at the transplant of fecal microbiota (FMT). As mentioned before, this is a transfer of microbiota from a donor to a recipient with the aim of restoring a normal population of microbic communities.

The question could be: "Doctor, why treat a single thing when the disease involves the entire ecosystem?".

Today we know that FMT is the best method for curing antibiotic-resistant *Clostridium difficile* infection. Many important diseases are related to a dysbiosis or in any case are associated with it and could benefit from FMT. One of the biggest problems is the standardization of the FMT protocols, starting from the apparently simple question: what do we mean by "healthy donor"?

Kidney and microbiomics

In physiological conditions, the predominance of symbiont ("good") bacteria, a healthy intestinal mucosa, the production of defensin, and the secretion of local IgA support the symbiosis between the host and its intestinal microbiota by blocking the growth of pathobiont (potentially "evil") bacteria within the intestinal lumen. A kidney disease can upset this equilibrium and trigger reciprocal metabolic changes: the progression of a chronic kidney disease toward an end-stage renal disease modifies the delicate balance between symbionts and pathobionts, thus causing an excessive growth of the latter (dysbiosis); we do know that bacterial metabolites, such as phenols, indoles, and amines, can contribute to the progression of an advanced chronic kidney disease and induce kidney injury, and that some of these metabolites, such as p-cresol and indoxyl sulfate, are associated with the diet–microbiota interaction. The close correlation between the structure of the microbic community and some

Metabolomics and Microbiomics. http://dx.doi.org/10.1016/B978-0-12-805305-8.00009-1

metabolites opens up the possibility of monitoring the metabolites themselves and manipulating the intestinal bacterial microflora that causes acute or chronic pathologies, as we will see in the next section.

Autism spectrum disorders and microbiomics

Autism spectrum disorders (ASD) deserve some reflections. They are permanent disorders of neurological development characterized by compromised verbal and nonverbal communication, a deficit in socialization, and repetitive behavior. The incidence of ASD has increased enormously: from 1 out of 2500 in 1980 to 1 out of 68 in 2014, with a ratio of 4 males to 1 female. In most cases, the cause has not been identified and there is noteworthy genetic heterogeneity. Monozygotic (identical) twins have a rate of concordance between 36% and 96%, dizygotic ones between 0% and 30%. The index of recurrence in siblings is ten times that of the overall population. In 1 family out of 10 there is more than one child with ASD. In ASD children we have observed an abnormal colonization of *Clostridia* spp. to the detriment of other bacteria such as *Bifidobacteria* spp. Even in our experience in the urine of these children we have found high concentrations of 3-(3-hydroxyphenyl)-3-hydroxypropanoic acid, which comes from the interaction between the prevailing presence of *Clostridia* spp. in the microbiota and diet. By changing the diet or altering the microbiota (using prebiotics, probiotics, symbiotics, or microbiota transplant) we can try to modulate the phenotypic expressivity of the disease. It is known that some ASD individuals benefit from diets without gluten and casein, with an improvement in cognition and socialization. Finally, there is a pilot study in which some ASD children received an FMT with 20 species of intestinal bacteria, with resulting clear improvements in speech, understanding, and clinical performance.

The strange case of the Caesarean section

The Caesarean section (CS) is the most frequent surgical procedure undergone by women all over the world and is reportedly increasing. Since 1985, the World Health Organization has recommended not to exceed the threshold of 15% of all deliveries. In reality, the percentage may depend on the level of the hospital structure, on the complexity of the cases treated, and the number of births managed. In any case, Italy has the highest percentage of CS in Europe (36.3% in 2013), whereas in northern European countries the percentages are much lower (14%). In some parts of Italy it is more frequent to give birth with CS than with spontaneous delivery. There is evidence to the effect that the microbiota of neonates delivered spontaneously is similar to the vaginal and gastrointestinal microbiota of the mother. The maternal vagina transmits bacteria that are useful in life outside the womb, such as *Lactobacillaceae*, which help to digest milk. On the contrary, the fecal microbiota of neonates delivered with CS is similar to that of the mother's skin. Children delivered with CS have in their intestines fewer lactobacilli and more hospital microorganisms, such as those from the wards, the pharmacy, and the storeroom. Different groups of bacteria are acquired early during the so-called period of "permissiveness" or "tolerability" and persist for a long time, to become dominant or subdominant, whereas others appear or disappear at a later time. Recent studies have found high interindividual variability in preterm neonates.

Mothers who have undergone CS are less likely to breastfeed their child and this deprives it of milk-oriented microbiota (MOM), which we spoke of before.

We are what we eat

Eating does not simply mean "nourishing oneself": the human diet is strongly influenced by cultural, economic, social, and environmental factors, which result in completely different lifestyles and diets. Even language is permeated by the more or less physical or mental relationship that every culture establishes with food. In the ancient world, the terms used in Greece and Rome to indicate a child came from different roots, a difference that indicated a different cultural perception of breastfeeding. The Latin word *filius* derives from an Indo-European root meaning to suck while the Greek word *yiòs* derives from an Indo-European root meaning to generate (from which also comes the English *son*). Among Italic peoples, the relationship was more physical, so much so that the term *nutrix* indicated the woman who breastfed and who concretely provided the nutriment needed by the child. On the contrary, in Greek culture, the act of nourishing a child, thus breastfeeding, recalled too closely the animal component of human beings and was looked down upon, so there is no precise synonym for *nutrix*. The equivalent term, *kourotrophos* (its gender is both male and female) generically indicated he or she who provided the means for raising children.

Food emphasizes our differences as peoples and ethnic groups, in our social standing and distinguishes us from others, but at the same time it can strengthen our sense of identity and belonging to a group through choice and exclusion, which sometimes reaches the status of food taboos. In Western societies today the choices of food are made on the basis of many different factors: we have the health foods of the vegetarians, food for meditation of oriental origin, junk food (the fast-food people), and the body-builder's diet of energizing beverages. These tendencies appear to be the result of a cultural operation (in the broadest sense of the term, which also includes the fashions of the moment), while the limits imposed by the need to use what our environment and climate offer, so decisive in the past, have now been overcome. It is becoming increasingly necessary to convince our societies of the benefits that can be obtained by consuming farm-to-fork food and promoting it. Although nature now appears easier to tame, in reality and in the light of new scientific knowledge, it is taking its revenge: our food preferences, choices, and taboos are dictated by nothing but our microbiota. Food choices are associated with different orientations of our microbiota which, in relation to our genome, produces different metabolites, which in turn shape us from infancy and control us: we are what we eat.

Bacteria that love chocolate and... Nobel laureates

I have decided to end this chapter with some curiosities on the consumption of chocolate, starting from this question: why are there "chocolateophyles" (those who love chocolate) and "chocolateophobes" (those who hate chocolate)? If I take a quick poll among those I know, the former group wins by a large majority. But even more interesting is to understand why chocolate lovers are divided between those who love dark chocolate and those who prefer milk chocolate. At a dinner, I received the answer from an illustrious gastroenterologist: he said it was all the fault of our intestinal bacteria, which decide our choices of food without our knowing this. Dark chocolate has a positive impact on our mood: 40 g of this chocolate every day for 2 weeks eaten by an adult has the effect of reducing the stress hormones.

But the most surprising thing is to discover if there is a correlation between chocolate consumption in a given country and the number of Nobel laureates per inhabitant. The study of this has provided an affirmative answer, and it is no coincidence that the country at the top of the list is Switzerland. Consumption of chocolate may in theory improve the cognitive functions not only of single individuals, but also of entire populations. In any case, it would appear that Nobel laureates are chocolate lovers, but that is not enough to win the prize. After all we have said about epigenetics, the real secret for winning a Nobel prize would appear to be to eat more chocolate! Does the winning of the prize depend on the bacteria in our intestines, on our enterotype? In any case, I must point out that this study has received much criticism from the standpoint of methodology: consumption was evaluated in the present, the effect on cognitive functions was measured in elderly people at risk of dementia, chocolate consumption by Nobel laureates was not assessed, countries such as Israel, with a low intake of chocolate and a high concentration of awards, were omitted. All this may have reduced the effect of the data presented.

Personalized medicine

10

The medicine of complexity

To illustrate the delicate, central, and crucial subject of complexity, I will use the example of neonatal asphyxia, both in animal models and human neonates. Some years ago, while studying the urinary metabolomics of newborn piglets that had undergone experimental asphyxia, our research team was greatly surprised to find that we could predict in all confidence whether they would die or not. What was most startling, and in a certain sense disquieting in the light of what emerges constantly in the questions I am asked at meetings when I speak of this, is that this prediction was made on the basis of the piglets' urine prior to resuscitation. The piglets had all been raised by the same farmer, the sows had all received the same controlled diet, the asphyxia model was the same, resuscitation was performed by the same resuscitator using the same technique and drugs. Some piglets died while others survived; of those that survived some returned to their basal conditions in less than 15 minutes after asystole and/or shock, others took as long as 2 hours, with insufficiency of various organs and apparatuses (lung, kidney, liver, and so on). The capacity to respond to and survive an extremely strong and acute stimulus, such as asphyxia, thus appeared to be an intrinsic property of the subject, practically independent of the application or nonapplication of treatment protocols: some newborn piglets were too fragile to cope with strong complications and died, whereas others were resilient and survived, with or without insufficiency of one or more organs. Why is this? It depends on their metabolism: their basal metabolome was quite different from the start. There is a strong basal interindividual variability from the metabolic and energetic standpoints which increases following a strong stimulus, such as asphyxia (but it could also be an infection or prolonged abstinence from food). This study has been confirmed by others on asphyxia and sepsis (a generalized infection of the organism), both experimentally and in human pathology.

But is all this ethical?

When I present these data I am frequently asked if this capability to predict and the consequent possible change in the physician's behavior is ethical. I was even asked during a discussion if metabolomics will replace ethics. The answer to the latter question evidently can only be negative; on the contrary, metabolomics will contribute to the assessing of phenomena more in depth and more precisely, which is to say in a personalized way, and this will help in making more appropriate ethical decisions for each single patient.

But is what we do today ethical? Today, we apply the same protocols to all our patients, a bit like dressing everyone with clothing of the same size. But we are quite

Metabolomics and Microbiomics. http://dx.doi.org/10.1016/B978-0-12-805305-8.00010-8

different one from the other. Some would not be able to button their jackets, others would have trousers that are too short, others still would have their shirt cuffs hanging below their hands. Applying this example to asphyxia, some patients die apparently from respiratory causes, others from heart complications, but today, excluding metabolomics and based on what we know, we treat them all in the same way, some too aggressively, others not aggressively enough. Some patients who die today could survive with personalized treatments. It is like going to the tailor for a custom fit. The path to take is to switch from protocols to personalized medicine, which basically means to accept a return to the past. The great poet T.S. Eliot wrote: "We shall not cease from exploration, and the end of all our exploring will be to arrive where we started and know the place for the first time." Let us go back to Hippocrates.

The hateful medical protocols

Contemporary medicine is strongly influenced, not to say obsessed, by health protocols. These are often necessary in countries with advanced health systems, even as points of reference for defensive medicine. In reality, by definition the protocols do not take into account (or only marginally so) the individuality of the single patient, a situation that would require a personalized, customized, made-to-measure diagnostic approach and treatment. Although protocols offer an unquestionable advantage compared to the anarchy of potentially arbitrary medical decisions, they do have their weak points. Firstly, they crystallize decisions and are thus not adaptable to single cases. It is often difficult to assess the appropriateness of a treatment or procedure in a single patient; secondly, they tend to relieve physicians of responsibilities: they can simply apply the guidelines and, for example, treat the glycemia of diabetic patients and not the diabetic patients themselves, who are unique and differ one from all the others; thirdly, the protocols tend to blur the physician–patient relationship because they do not necessarily respond to patients' needs, but especially to the exigencies of defensive medicine. When they are approved and published, the protocols come into the world already outdated and obsolete by definition after a long, hard-fought gestation, and in the end they have a short life. The protocol may represent an emotional cushion (a sort of avoidance mechanism) between physician and patient, in particular when the situation is quite complex and critical; the protocol can even become, or be used, as an authentic barrier. Is it not better to return to bedside medicine, honoring our patients and their histories? I hope that you noticed the verb "to honor" taken from Rita Charon, who has systematized narrative-based medicine.

Usain Bolt and the secret language of destiny

On the subject of wide-ranging interindividual variability, we can quote Montaigne: "There is a greater difference between one man and another than between a man and an animal". We are living beings of an enormous and infinite complexity. A further proof of what is said above is the inestimable variety of the human face: no face is the same as the face of another. This characteristic distinguishes the human species. Only an awareness of this complexity and the biological variability associated with it will increase our knowledge and enable us to follow a proper diet and treat our patients in a personalized way on the basis of their individual differences, and not by dealing with a hypothetical "average patient." Moreover, one of the most frequent complaints of patients, especially in certain fields such as oncology, is that of being placed in a protocol that has no consideration for their individuality. Two examples: the first has to do with sport. Let's say that we are watching the final one hundred-meter dash at the Olympic Games. On their blocks all the participants seem quite similar, but only one of them is Usain Bolt and almost certainly he will win the gold medal. There is another sprinter (let's call him Anonymous) who got to the finals by luck because in his semifinal the fastest runner tripped and fell. Even though Anonymous has been training for several hours every day for years, he will never succeed in winning a gold medal at the Olympics because his basal metabolism is much different from that of Bolt's. The difference, already clear, between the basal metabolisms of Anonymous and Bolt, will show up spectacularly during the enormous stress of the one hundred-meter race. And what if we put a man from the street, let's say the writer of this book, in the race? No answer is required. So why is it that we want, even expect, all patients to be like Bolt and react like champions to the massive psychophysical stress of disease?

The second example is strictly medical: we have three neonates born at the regular end of pregnancy but who suffered asphyxia at birth (from the Greek *asphyxía* "arrest of the pulse," composed of the privative *alpha* and *sphýxis* "pulsation") with oxygen deprivation. This can take place at term even when up to the final moment everything has proceeded regularly. The three babies appear before the neonatologist with the same calling card, but unfortunately with a low starting score (without going into detail, this means a low Apgar score, used to assess a neonate's vitality and the efficiency of its primary functions), with the same alterations that show up in the laboratory tests (in this case the acid–base balance) and with the same neurological symptoms that place them in the hypothermal protocol, a procedure that cools the brain for 3 days to limit brain damage caused by a lack of oxygenation. From the medical standpoint, the three babies are the same: same history, same Apgar score at birth, same laboratory results in use in the units today, same treatment protocol. Now let us examine the results. Unfortunately, patient A dies after 48 h, one day before completing the hypothermal treatment. Patient B worsens clinically, with kidney and liver insufficiency: he survives, but on dismissal he clinically presents evident neurological damage. Patient C survives and on dismissal seems to be a normal baby dismissed from the nursery and not from the neonatal intensive care unit. The instrumental tests show injury, but clinically the signs are almost imperceptible. Why is this? How is all this possible if the physicians made no mistakes, they followed the international and national protocols competently, carefully, and with no negligence? The answer lies in the babies' extraordinary interindividual variability. In the three, the basal metabolome is quite different, and it differs even more as a response to asphyxia at birth. We would all like the three babies to be like Bolt, but it turns out that one is like Bolt, one is like the finalist by chance, and the other is like the man in the street. Neonate A was fragile, C resilient, and B halfway between. What is the physiopathological explanation revealed by the metabolome and metabolites? Neonate A was unable to transform resuscitation into energy: he had used up all of his energy, which was scarce even at the beginning, and he was burnt out much too soon after the strong stimulus. No physician would have been able to save the child by applying the national and international protocols in use today. Nowadays, it is quite probable that the doctor would be accused of malpractice for failing to save neonate A's life or the brain of neonate B. But the cause is also, and sometimes exclusively, in the intrinsic diversity of the patients and/or the differences in their lives in the womb (obviously excluding documented cases of malpractice, carelessness, and negligence). Only through an understanding and better definition of the state of health of each individual, his or her resilience or fragility, will medicine be in a position to respond in a personalized way and not approximately, or only epidemiologically, to the problems of human health.

Following is the example of nephrons to illustrate the key concept of interindividual variability. I will then go on to explore the new directions of research in the following chapter.

The "treasure" of our nephrons

Here I will speak of the kidney, an organ to which I have dedicated many years of study. The person who aroused this interest in me is Gordon Avery, the author of one of the fundamental texts on neonatology. When I was quite young, I attended one of his outstanding lectures. His answer to one of my questions was that neonatologists know everything about the brain, heart, and lungs, but nothing about the kidney. He urged me to study the kidney and that is what I have done.

In the kidney, the nephrons are the functional unit formed by the glomerule (the "prima ballerina" of the kidney), the tubule (composed of the most highly specialized cells of the organism after the cerebral neurons), blood vessels, and finally the interstice in the middle. The nephrons are hard workers who toil in silence 24 h a day and are extremely important for the entire organism since they remove a part of the body's wastes. They are an authentic treasure (in my lectures I pronounce the word in the scratchy voice of Gollum in the film version of *The Lord of The Rings* so that my medical students keep awake even in classes just after the lunch break). If I ask my students how many nephrons we have, most of them look down at the floor. In reality, the question is not so easy, since very few studies have been performed in this direction. The answer is that the number varies from 200,000 to 2.5 million and this is considered normal for a person. Do you recall the idea of our extraordinary interindividual variability? Why is there such a big difference? In normal conditions nobody is aware of the difference since our organism has built-in redundancy. We have many more nephrons than we actually need. It is like having a Formula 1 car and using it only in the city at 30 miles an hour: I don't need all the power of that engine in normal conditions. Let's imagine three other neonates (but we could also say three people in general), the first with 200,000 nephrons, the second with a million, and the third with 2.5 million. If I perform laboratory tests and the creatininemia is normal, the kidney function is considered regular. If there has been acute kidney injury that "kills" 100,000 nephrons as the result of hyponutrition, hypo-oxygenation or a drug that is harmful to the kidney. In the first neonate this corresponds to the loss of half its nephrons. This means that in the pediatric age he will have chronic renal insufficiency. In the second neonate the damage corresponds to one-tenth of its nephrons so there may be long-term damage in adulthood or old age. For the third neonate this loss is of little consequence, negligible, and insignificant in the short and long term.

To conclude, besides differences between one hospital and another and from one doctor to another in the more or less correct application of the protocols, we must also take into account an element that is often underestimated: that of the basal variability between one patient and another and their variability after treatment.

The new approaches of research

How the course of research has changed

Barker was a fundamental linchpin, a sort of "black node" of scale-free networks. He taught us the need to further develop our knowledge of what takes place in the fetus and in the first 1000 days of life from conception, since this period represents a window both of vulnerability and opportunity, with repercussions on a person's entire life, as I pointed out in the chapter devoted to perinatal programming.

However, the concept of perinatal programming (or developmental origins of health and disease) is still little known, too little in consideration of its real importance. Suffice it to say that specialists still do not ask their adult patients the essential question: "How much did you weigh at birth?"

Unexpected similarities

In our immunohistochemistry research performed in Cagliari together with the pathologist Gavino Faa and his team, unexpected similarities emerged between mechanisms essential in fetal life for the morphological and functional growth of the embryo and tumoral processes. A slogan that I often repeat to my co-workers is: "Substance is what counts, but form informs." Once again structure and function. Recovery from a tumor may be inborn in our very nature (substance), since the metabolic pathways of the tumor may be the same as those of the embryo (form, or what we can observe). From our studies, especially those on the fetus kidney and liver, it has emerged that these organs express high concentrations of markers that are also typical of highly malignant tumors, such as WT1 and MUC1 for the kidney and Glypican 3 for the liver. WT1 stands for *Wilms tumor 1*, and is a typical pathognomonic marker of Wilms tumor (or nephroblastoma), a child's kidney tumor. MUC1 is expressed by tumors of the kidney and other organs of the adult such as the breast, ovary, pancreas, colon, and lung. Glypican 3 is the biomarker of hepatocarcinoma. I will spare you the list of other markers because what I want to emphasize is that these substances are essential for the development of organs during fetal life and afterwards are deactivated. At a certain point, for reasons that are still not clear, they are reactivated and cause tumors.

What is the objective of a tumor? In all likelihood it is that of growing as quickly as possible to conquer the entire organism along the most ergonomic pathway. So it uses the embryonic systems, which ensure growth at a rate far above what occurs in other situations. The tumor uses the embryonic systems also because they enable it to grow

Metabolomics and Microbiomics. http://dx.doi.org/10.1016/B978-0-12-805305-8.00011-X

in an organism "against everything and everybody." At no other time in life is growth as fast as it is in the prenatal period, not even in adolescence. Some factors can act as accelerators, others as brakes, but not all brakes are the same: how can we identify the best ones? If during embryonic life the organism was capable of blocking a metabolic pathway at an early time, why couldn't the same thing take place at a later time for tumors, which follow the same metabolic pathway?

Kidney regenerative medicine

In speaking of regenerative medicine, once again I offer the example of the kidney. Up to now, all researchers have focused on what inhibits nephrogenesis. Why not instead concentrate on what favors nephrogenesis in the preterm neonate? The organism maintains a delicate balance between protective and aggressive factors in all its components. In the case of the kidney we can imagine that we can influence nephrogenesis both by releasing the brake to reduce or inhibit programmed cell death (apoptosis), and by acting on the accelerator to favor the processes of cell replication (mytosis). This concept is illustrated in Fig. 11.1. The action of the two pedals: brake, and accelerator, modulated by different physiological and pathological factors, may prolong the period of nephrogenesis and increase the production of nephrons at the right time, or possibly both options.

The kidney also contains stem cells. Better knowledge of the stages in the maturation of these cells will enable us to face the future with more weapons than in the past to improve the prognosis of patients affected by kidney insufficiency, both acute and chronic.

The new frontiers of research

Research must go in the direction of new ambitious goals, but must also consider past results from a novel standpoint. As Feynman said, the winning insight is not only to have new ideas, but also to arrive at better results with those we already have.

I believe that in the neonatological and pediatric ambits we face important challenges that we must accept and win. I list them in Table 11.1, together with the fields of scientific research and the objectives most focused on the social plane, to emphasize that science cannot consider itself outside the historical and social contexts in which it develops. Socioeconomic reality necessarily impacts on and conditions both research and the management of public and private health systems. It is no coincidence that the first two entries in the table emphasize two crucial issues: can we improve neonatal and pediatric survival worldwide, in countries with low social and health conditions and scarcity of funds? And even in countries with good health care but in a context of drastic cuts in resources destined for health services, is it possible to provide high-technology medicine? As is said, health is priceless, but health care has a cost. In all probability there is no single answer, but the need for a global strategic perspective.

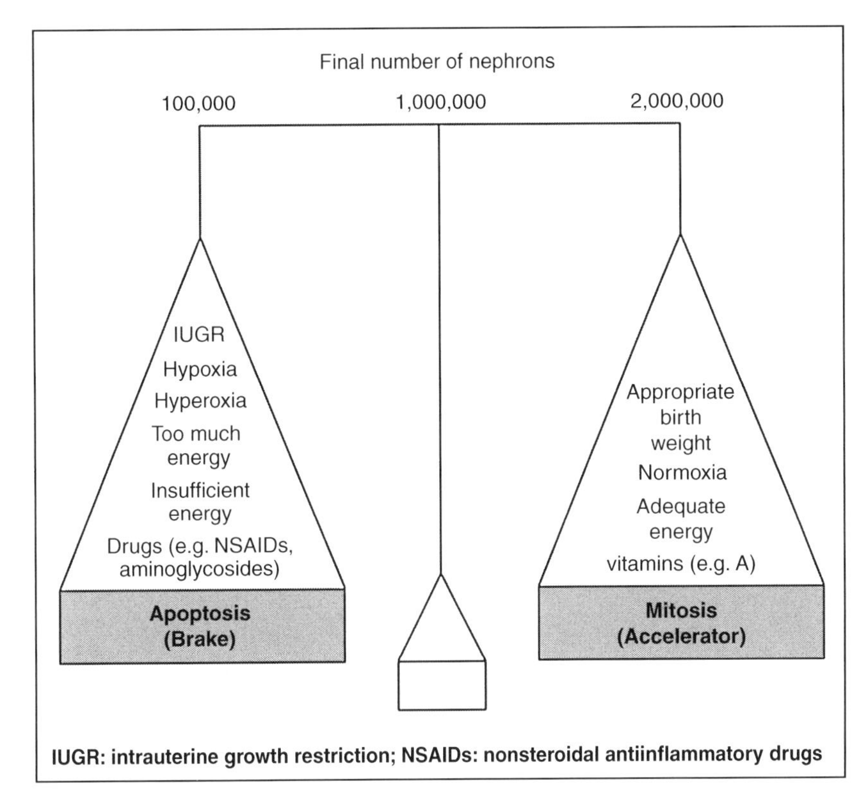

Figure 11.1 In the case of the kidney we can hypothesize that we can influence nephrogenesis both by releasing the brake to reduce or block programmed cell death (apoptosis), and by pressing on the accelerator to favor the processes of cell replication (mitosis). The action of the two pedals, brake and accelerator, modulated by different physiological and pathological factors, may prolong the period of nephrogenesis and increase the production of nephrons at the right time in the window of opportunity, or possibly both. The fact that on the basis of these factors, the scales will tip to the right or left thus determining the final number of nephrons present at birth and, consequently, a person's "renal destiny." Modified from Fanni et al. (2012a).

The need to allocate important resources to the prenatal, neonatal, and pediatric periods derives from evidence that most perinatal and pediatric deaths can be prevented, and that prevention of adult diseases must begin as early as possible, even in the womb or the first period of life. These are the times when a preventive measure has the maximum and most long-lasting effect and, in all likelihood, the lowest cost.

We must have efficient research organizations and networks. It is of the utmost importance to decide on the right moves to avoid wasting time and resources and plan, perform, and analyze the research we perform, in which we must aim at Edison's, but especially Pasteur's, quadrants (Fig. 11.2). In any case, we must satisfy the needs of health care to produce fallout as quickly as possible in the front lines of our health structure battlefields. The issue that is most felt is not so much the discovery

Table 11.1 New objectives and directions in neonatology and pediatrics

- Reduction of neonatal and pediatric mortality, especially in developing countries
- Reduction of the cost of health care while at the same time maintaining a high quality of treatment
- Prevention of long-term complications in connection with perinatal programming
- Study of the long-term consequences of the kind of delivery (spontaneous or Caesarean)
- In-depth study of gender medicine in perinatology
- Implementation of neuroprotective measures in perinatal medicine
- Noninvasive sampling and early diagnosis of pathologies
- Development of nutrimetabolomics and pharmacometabolomics
- Immunohistochemical study of embryo-fetal tissues and regenerative medicine
- Advanced simulation
- Implementation of medical humanities
- Further development of narrative-based medicine
- Knowledge and understanding of the similarities and differences between studies on animals and humans

Modified from Fanos (2012b).

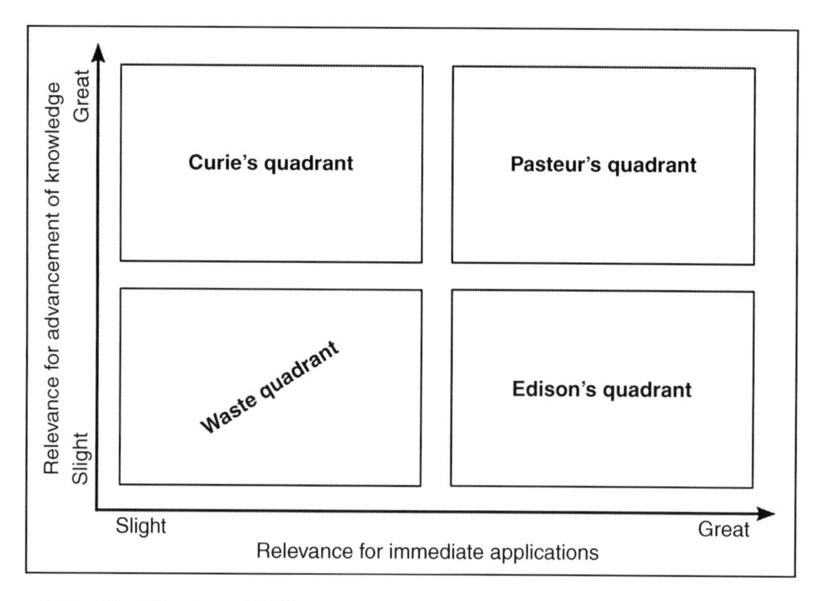

Figure 11.2 Classification of different research categories in four quadrants. We have kept the classification of Chalmers and coworkers as concerns the Curie and waste quadrants (on the left in the figure) and that of Stokes for the Pasteur and Edison quadrants (on the right).
Modified from Stokes (1997) and Chalmers et al. (2014).

and exploitation of talents, but above all of the so-called "Valley of Death," the wide gap, the immense vacuum, between basic research and its clinical applications. As the often-cited Altman prophetically said: "We need less research, better research, and research done for the right reason".

The most important branches of research are those needed for the advancement of knowledge, but which are also capable of responding to important practical questions, which is to say those that fall in Pasteur's quadrant. I have already discussed or mentioned some of these branches on the pages of this book. I have spoken at length on how a child is delivered, the impact of delivery on the intestinal microbiota and the long-term implications of perinatal programming. I have also spoken of nutrimetabolomics and pharmacometabolomics, of immunohistochemistry and regenerative medicine, as well as the need to find holistic, early, discriminating, and predictive biomarkers. The description of advanced simulation is beyond the scope of this text and readers are referred to specific articles on the subject. Finally, in the next chapter I will discuss medical humanities and narrative-based medicine. Here I conclude by emphasizing two points: gender medicine in perinatology and the similarities and differences between animals and humans in the field of the science of healing.

Gender medicine in perinatology

Gender medicine is based on the fact that since males and females are biologically diverse, they react differently to diseases and respond differently to treatment. Although this seems obvious, in reality little attention has been focused on this subject in the adult and even less in the early years of life. Owing to a lack of sufficient data on the perinatal period, I will cite examples in adulthood. These examples are in any case fairly valid in the perinatal context because today we know that the different periods in life are in reality closely interconnected.

- Thirty-eight percent of women who have had a heart attack die within 1 year compared to 25% of men. Much of the information we have on drugs that act at the cardiovascular level has been obtained from studies involving males only; the routine application of these studies to the female universe leads to negative consequences, since drugs are less effective and more toxic in females.
- Generally speaking, women with diabetes have a poorer quality of life and a shorter life expectancy than diabetic men.
- The prevalence of asthma has a twofold tendency: before puberty males are affected twice as often as females. After sexual maturity this difference disappears with hormonal changes.
- A woman's lungs, even those of nonsmokers, are more exposed to lung cancer, especially in their fertile period.
- Parkinson's disease is up to twice as frequent in men, but women are more severely affected.
- There are differences concerning peptic ulcers and sensitivity to pain.
- The impact of alcohol on the male and female metabolisms is quite different, independent of weight.

We can speak of a "woman's paradox": women live longer than men, but they are more often ill, they rely more on the health system, and take more drugs, with a frequency of side effects one and a half to one and three-quarter times higher than men but in practice the patient information leaflets that accompany drugs make no mention of gender differences.

In the neonatal period, the concept of gender medicine has been known for at least 40 years, since Naeye spoke of the "male disadvantage." In prenatal life there is a struggle between paternal genes that "root" for the fetus and want to keep it alive and thriving at the expense of the mother, and maternal genes that are on the mother's side and want to keep her alive to have other children, even at the expense of the pregnancy in course. The mother must modify and lower her immunitary defenses to keep the fetus in the womb. For the maternal organism it is easier to "tolerate" a female fetus (XX chromosomes like the mother), than a male fetus (XY chromosomes). In reality, more male neonates die, more males are born preterm at higher risk of developing diseases such as neonatal respiratory distress, necrotizing enterocolitis, chronic lung disease (also called bronchopulmonary dysplasia) and cerebral hemorrhage. Although these data appear to be a bit out of date, every neonatologist knows that there is some truth in this.

Animals and the science of healing

At the end of this chapter on the new branches of research I cannot omit a reflection on the fact that although many experimental data have been obtained through studies on animals, which do not always completely correspond to humans, certain animal diseases can shed light on human diseases and the science of healing.

Our contiguity with the animal world struck me clearly while reading the book *Zoobiquity*, with the intriguing subtitle: *The Astonishing Connection Between Human and Animal Health*. The book contains brilliant ideas and curious definitions, like that of the physician as the veterinarian of a single animal species. The beginning catches the reader's attention: the author, the cardiologist Barbara Natterson-Horowitz, was called to the Los Angeles zoo for a consultation on an emperor tamarin monkey with a heart issue, a truly unusual patient for her. This is a primate that owes its name to its long moustache like the kind in fashion during the reign of Emperor Wilhelm II of Germany. While examining the animal, she learned from the veterinarian that wild animals can die from a form of sudden cardiac arrest in connection with an extremely stressful event. The phenomenon is known as *capture myopathy* because when an animal is captured it has an enormous discharge of adrenalin capable of paralyzing the heart. But the cardiologist knew this phenomenon well for having found it in some patients. Faced with severe stress, this can also take place in humans and in this case it takes the name of *Takotsubo cardiomyopathy*. It is a fairly rare event: in the overall population the ratio is 1:36,000 and between males and females about 1:3. The most typical patient is a woman in postmenopause with no significant cardiovascular risk factors. In the anamnesis of many affected persons we find prolonged emotional distress: we also speak of the broken heart syndrome. It is often the case of persons who see a loved one die suddenly before their eyes. Different names, the same physiopathology, the same problem. The veterinarian knew about it 20 years before the cardiologist!

Our closeness to the animal world is truly noteworthy: humans are incredibly similar to animals, but we must remember that we are also incredibly different from them, and this is no contradiction. This fact must be kept in mind when we assess results obtained on experimental animals and we must be careful not to apply them uncritically to humans. Here is another example from the kidney. As we have seen, nephrogenesis in humans is completed in the 35th or 36th week of gestational age. Nephrogenesis thus ends in the prenatal period and this is the same for mice and sheep. The animal whose kidney is most like that of humans is the pig, which however, like rats and rabbits, completes its nephrogenesis in the postnatal period. So to study the kidney we are faced with a doubt: if we consider the period of maturation we should choose the mouse, but mouse kidneys have few, very long tubular structures, while in the human kidney evolution has privileged a large number of shorter tubules. If we want to study the kidney structure, we should choose the pig, but in this animal nephrogenesis continues after birth. The pig kidney, although similar, is much more rudimentary and simpler than that of humans. Finally, the human kidney is larger than that of mice, rats, rabbits, and also cats and dogs, while it is smaller than that of bears, tigers, pigs, and lions (in order of increasing size). We must not forget the similarities, but the same is true for the differences. No model is perfect, no single one is the same as the human kidney. So we have to compromise. We have studied animals to such an extent that if a mouse is ill, we have all the information on how to treat it. The question is: what if a person is ill? In this case the doubts are many, we have many more things to discover and understand: we are faced with an extraordinary complexity.

The medical humanities

12

The meeting between physician and patient

At the beginning of this book I wanted it to be clear that while being dazzled by technology, we must never lose sight of the very special significance of the relationship between physician and patient.

The risk of the medicine of the future will in fact be that of presenting itself, or being perceived, as muscle-bound and infallible, on the verge of the miraculous. However, medicine cannot be detached from patients and their needs.

This chapter is thus devoted, albeit briefly, to an exploration of the characteristics of the doctor–patient relationship, a quite special aspect of communication.

Medical humanities: in the halfway land of complexity

Speaking of this subject, I have decided to start from the Greek physician Jason, engraved on a stele I saw while visiting the *Hygieia* exhibition at the Cycladic Art Museum of Athens in December of 2014. The subtitle was *Health, Illness, Treatment from Homer to Galen*. The marble relief, found in Athens and now at the British Museum in London, shows the physician Jason who is calmly and unhurriedly examining one of his patients, who shows evident signs of illness. With great realism, the sculptor caught the tension in the eye contact between doctor and patient. It is a time of anxiety for the patient, who looks toward the doctor to be reassured. At the same time Jason is concentrating and bringing into play all his knowledge and experience to understand the signs and symptoms leading to a diagnosis. The two are looking each other in the eye, which at that time was considered the mirror of the soul. Medical humanities are all here, and at this point I could end the paragraph.

But let us examine the world today: physicians are sometimes in a hurry, being chased by statistics on the timing of examinations and evaluated on the basis of the economic turnover of their workplaces, which today in Italy are called *aziende* (enterprises), a truly obnoxious term for a hospital. And what can we say of the Dantesque circles of Hell that are our emergency rooms? Anamnesis, physical examination, differential diagnosis, treatment and explanation of the cure in the shortest possible time. Unfortunately, in some cases something must be skipped, and that something is the anamnesis. What a mistake! Patients today complain of this black hole and may turn to alternative medicine, in which the doctor devotes much more time to the anamnesis. Or they search the Internet and become their own doctors. An oncologist from Milan,

Metabolomics and Microbiomics. http://dx.doi.org/10.1016/B978-0-12-805305-8.00012-1

Table 12.1 Problems sometimes present in today's hospitals

Problems
• Super-specialization
• Technicism
• Super-professionalization
• Hyper-medicalization
• Insensitivity to personal and cultural values
• Much cure and little care
• Too much science
• Callous treatment of patients
• Deficit in verbal and nonverbal communication
• Superficiality and arrogance

who had had a stroke, observed that what was needed in medical schools was another course: that of humanity.

Naisbitt wrote that technology is a precious instrument, a great resource; but in medicine "high tech," which is to say super technology, must not lead us to forget "high touch," that is, human contact.

Some of the problems that today's hospitals are called upon to cope with are presented briefly in Table 12.1. This table, in modified form, is taken from a work by Edmund Pellegrino in 1979. More than 35 years have gone by, but it appears that nothing has changed: we could speak of "misadventures in the hospital."

The improper use of *evidence-based medicine* divorced from an in-depth analysis of the themes of relationship and communication has progressively led physicians to focus their attention on the disease and not on the patient. Health providers take into consideration a diseased body or a part of it rather than the person who is ill. This attitude has brought about a process of alienation and in many cases the physician–patient relationship has become an experience that lacks authenticity.

The variegated movement of medical humanities represents a reaction to this state of affairs and it is an extraordinary opportunity for medicine to return to its profound human dimension, as it was in Jason's time. As I mentioned before, we can treat patients more effectively by *Honoring the Stories of Illness* to cite the title of a book by Rita Charon, the founder of narrative-based medicine. By honoring these stories we can extend our horizons and expand the narrow vision of physicians who focus on the biological disease by encouraging them to consider the emotional, social, and family needs of their patients.

Before speaking about narrative-based medicine, I would like to mention the experience in our neonatal intensive care unit, neonatal pathology and nursery in Cagliari, Italy. We have a consultant working with us who has a degree in philosophy. Together with him and all my colleagues, the obstetricians, psychologists, and psychiatrists, we have for years been implementing a series of programs that range from screening of postpartum depression for the mothers of neonates in the neonatal intensive care unit, communication to parents of the first diagnosis of the Down syndrome, therapeutic education, narrative medicine and consultancy of an ethical nature (for example, what decisions to make if a baby is born live but with a pathology incompatible with life).

I believe it is appropriate to close this paragraph with two quotes, the first by Cosmacini and the second by Martignoni:

- "Medicine is not a science, it is a practice based on science that operates in a world of values. It is, in other words—in the Hippocratic sense of *techne*—a technique endowed with a knowledge of its own, cognitive and evaluative, which differs from other techniques because its object is a subject: man.";
- "They [the medical humanities] are not citizens of the terra firma (...) but are navigators with no fixed residence amidst the islands of an archipelago that obliges them to speak many dialects and experiment with many practices open to the interlacing and intrigues of complexity."

Isn't it singular to find the same complexity of the biological world in the world of relations?

Narrative-based medicine

Narrative-based medicine impacts on the management of patients. It is a model of empathy, reflection, professional conduct, and reliance in the doctor–patient relationship, one that requires great attention in listening to what patients have to say. Instead of approaching all patients in the same way, asking them only for information needed to fill in a form, narrative-based medicine enables the physician to understand what patients want to say and how they want to say it. We could say it allows a holistic and personalized approach.

With narrative-based medicine we can: (1) understand the version within the pragmatics of communication; (2) establish a therapeutic alliance with the patient; (3) share decisions with the patient. Concerning point (3), in the past, the paternalistic regime (which we saw at the beginning of this book, the catch phrase of which is "doctors know better") was in force: today, we have the patient's informed consent (this obviously must be truly informed), but tomorrow we will have shared decision making. Not everything is this simple and problems arise if there is disagreement between doctor and patient, between doctors and nurses and, in neonatology and pediatrics, between parents and doctors or even between the parents (when the mother and father do not agree or they change their minds about decisions previously shared). Physicians cannot provide information, numbers and percentages without keeping in mind the context of these patients and their families. As T.S. Eliot said: "Where is the wisdom we have lost in knowledge? Where is the knowledge we have lost in information?"

You will have noticed that I have filled this book with questions as well as quotes. I agree with Neil Postman and Charles Weingartner when they say: "Once you have learned how to ask questions—relevant and appropriate and substantial questions—you have learned how to learn and no one can keep you from learning, whatever you want or need to know."

Einstein said: "The important thing is not to stop questioning."

Scientific debate in the Internet age

Between evidence-based medicine and medicine-based evidence

In all medical fields it is important to encourage an up-to-date multidisciplinary approach, one that is also "tailored" to clinical issues. In Table 13.1 we see in condensed form some of the features of the medicine we would like to have, perhaps a utopia, but not so far from reality and in any case a goal to work toward. The alternative could be a discouraging medicine based on impressions, arrogance, and considerations of an exclusively economic nature, a medicine that could be perceived as strong, "muscle-bound," if not infallible, but in reality approximate and inattentive to the needs of the single patient.

We can use a short formula:

$$FM = EBM + MBE + OBM + NBM$$

where FM is the future of medicine, EBM is evidence-based medicine, MBE is medicine-based evidence (not a play on words: it is evidence based on clinical experience, based on what we know and have learned in practicing our profession day by day "in the front lines"), OBM is -omics-based medicine (medicine based on the -omics technologies), and NBM is narrative-based medicine (which means attributing great importance to communication and the medical humanities).

In scientific debate, however, on reading articles published in the leading journals and attending important congresses in the different medical disciplines, we have the impression of being pulled in different directions by opposing forces, at least apparently: in one direction by personalized medicine and in the other by meta-analyses, the statistical analyses of data collected in different studies on the same subject. Meta-analyses have been, are, and will always be important in medicine. They have led to enormous advances in the evolution of science, and yet we can say that the meta-analysis is not everything and that in it the absence of evidence is not the evidence of absence, which is to say that the lack of proof that a drug works is not proof that it does not work. Again, this is not a play on words. The articles selected for the meta-analysis are often few compared to those initially taken into consideration. Does this mean that the discarded works were a waste of time? This is obviously not the case because they have supplied elements and useful data in clinical practice, even without the meta-analysis. Someone has said as a provocation that meta-analyses serve only to demonstrate what one wishes to demonstrate. The history of medicine has progressed even without evidence-based medicine. In reality, meta-analyses often group together

Metabolomics and Microbiomics. http://dx.doi.org/10.1016/B978-0-12-805305-8.00013-3

Table 13.1 The medicine to promote

• Patient-oriented medicine
• Family-oriented medicine
• Evidence-based medicine
• Medicine based on ethics
• Medicine based on experience
• Medicine based on excellence

in the same basket pears, apples, oranges, and bananas, and what comes out is a fruit salad. What would be useful are meta-analyses of wide-ranging studies performed on large numbers of patients and conducted in many centers with uniform procedures, and meta-analyses that take into consideration not aggregated data, but data on the single individuals that participate in them.

Patients as Internauts

Millions of people use the Internet for information on their health. In Italy, one search out of five on the web has to do with health. This number is constantly increasing and is similar to data in other countries. On this subject I can propose some food for thought:

1. physicians and health professionals should pay more attention to the time that patients spend in front of their computers in search of medical information and should ask themselves why they do this;
2. it is useful to follow the trends of patients to gain an understanding of what is taking place;
3. it is important to avoid an overload of information;
4. medical quackery must be avoided.

Here is one example on the third point. In an article published in *Science*, the authors calculated that up to 2010, about 129 million books had been published and Google had scanned more than 15 million. Of the many data presented in the article, an interesting one was that if we tried to read only the books printed in the year 2000 at a reasonable speed of 200 words a minute, with no breaks for eating or sleeping, it would take us 80 years. The sequence of letters would be 1000 times longer than the human genome: if we wrote them in a single line it would make over ten round trips to the moon. What I mean is that the information available today is overwhelming, really too much, and we must know what, and how, to choose.

This is true not only for researchers and physicians but also for patients (the fourth point). Of all that we find in the Internet on the subject of medicine, only a small part is written by experts in the field. The collateral effect is medical quackery, medical babble that has nothing to do with medicine and borders on a scam, not lastly because we do not know who is responsible and for what. One of the consequences is the diffusion of false information that feeds on itself and appears to be true because it is repeated and reappears in continuation, but things do not come true only because they have been written or seen on television. Luciano Onder, a scientific journalist and television host, has said: "In its humble way, good information contributes to good medicine. Bad information aggravates problems and damages those who listen to it."

The question is why physicians do not communicate more with the general public or communicate badly. To divulge science in general and medicine specifically, Piero Angela, for many years the host of scientific programs on Italian television, states: "It is not enough to explain clearly: we must use a language that is understandable even by those who are not familiar with the subjects presented."

But the voice of physicians is often lost amidst the many others of different provenance and competence that create more confusion than information. Mario Perniola, in his essay-pamphlet provocatively entitled *Contro la comunicazione* (Against communication), holds that mass communication is capable of transforming inconclusiveness, confusion and recantation from a source of weakness, as it is in reality, into elements of strength that enable something to stay in the limelight for as long as possible. These elements apply a falsely democratic varnish on popular science so that everyone has the impression of being an expert and thus able to participate in the cognitive process on the same level. The term *democratainment* has been proposed to indicate this phenomenon, which is well known and studied by communications professionals. This is a deviation of communication that turns information in the direction of entertainment. Scientific communication conducted in this way has little to do with science and much to do with entertainment. The subtracting of space from the authoritative mediation of professionals in any sector of human activity and in any context does not go in the direction of clarity and reliability of information that emerges from it. If this is true in every sector, from art to economics, from law to politics, it is also true in the sciences and in particular medicine, the sector we are dealing with here. It is proper and legitimate for all to express their opinions, but not all of them belong on the same plane: some are professional, others motivated by interest in the subject and still others by momentary emotions. A lack of order and method detracts significance from communication: what is important is knowing who is saying what, for the just and legitimate protection of the public at large.

Messages to take home

14

In this final chapter I will try to systematize and summarize what I have said in the previous ones, starting from metabolomics, which is rapidly becoming the new clinical chemistry. This technology enables us to identify and quantify at the same time thousands of metabolites in the organism's different liquids and it thus represents a new holistic, sensitive, and predictive approach to diagnosis. Many results are now available and some of its most important applications are represented by pharmacometabolomics and nutrimetabolomics.

In the same way, giant steps forward have been taken in identifying the bacteria that live in our intestine and their hierarchical relationships (sociomicrobiology) and our three major enterotypes have been identified. The key message is that in the presence of a genetic predisposition only the interaction of a certain kind of diet with a certain type of intestinal microbiota produces metabolites that pass into the bloodstream and modulate the structure and function of our organs, including the brain. Many illnesses derive from an unbalance induced by an inappropriate diet that leads to dysbiosis, that is, the abnormal colonization by an intestinal bacterial species that outweighs the others. The intestinal microbiota interacts with the immune system and orients it. These forces are at work throughout our lives, but they are of the utmost importance during time windows when our development, both fetal and in the first years of our lives, is at risk.

There is a perinatal programming of organs that explains the consequences, even in the medium or long term, of early events during fetal development. Much of what takes place in perinatal programming has to do with energy processes and the enlisting of energy from different sources of the cell.

The combination of the factors mentioned previously explains why the improvement and optimization of nutrition play the leading role in maintaining a person's state of health, in preventing diseases and their complications, and in expressing the response of all of us to the different pathologies. Furthermore, the rebalancing of the intestinal microbiota with prebiotics, probiotics, and symbiotics (prebiotics plus probiotics), as well as with the transplant of intestinal microbiota with bacteria from a healthy subject, will lead to an improvement in our state of health and, above all, to its consolidation. The gut microbiota must be known (what enterotype are we?), protected, and respected and we must avoid disproportionate treatments with antibiotics. Even diseases that have reached epidemic proportions, such as obesity, may originate from the intake of antibiotics, especially when not called for, in critical moments of development as in the pre- and perinatal periods.

We are learning, not without difficulty, that a person's life must be seen as a continuum from prenatal life (even before conception!) to that of the adult, and in this journey the most important period is the time spent in the womb. In this realm every

Metabolomics and Microbiomics. http://dx.doi.org/10.1016/B978-0-12-805305-8.00014-5

effort at prevention—in defense of the expectant mother, the fetus, the neonate and the child—will produce better and longer-lasting results compared to those in every other period of life and will cost much less from the economic standpoint. We can do much by changing the epigenetic mechanisms in our favor: if we cannot change our DNA, we can modify its expression. DNA is not our destiny! We cannot change our past, but we can change our future.

Another key concept to remember in every situation is our extraordinary basal interindividual variability. This wide variability (that swings from fragility to resilience, to antifragility) is enormously accentuated after acute stimuli such as neonatal asphyxia, sepsis, or prolonged abstinence from food. Of this boundless interindividual variability, physicians and health providers in general are not sufficiently aware and even less is society in general, the mass media, and the magistrates called upon to judge cases of presumed malpractice.

This variability is further proof of the immense complexity of biological systems, faced with which we cannot but be humble. The most important obstacle to a real understanding of the sense of science is not so much our ignorance, but our knowledge, or better still, what we think we know, the preconceived ideas we have developed about things. It is true that we have taken many steps forward and gained far more knowledge than in the past, but this knowledge is fragmentary and not yet consolidated in a more complex and integrated system.

In the case of medicine, only the application of holistic technologies such as metabolomics and microbiomics will provide an image of the incredible complexity of every individual and capture his or her distinct and unique metabolic fingerprint and the pathways from a state of health to one of illness and from the latter to recovery. Today, these technologies are extremely complex, costly, and available in few places. But it is not difficult to foresee an explosion in research on metabolomic applications in the next 3–5 years, which will lead to the production of simple, cheap, and ergonomic instruments, such as kits for the metabolomic analysis of urine in different pathologies. This will take us from pure research to the patient's bedside, thus providing an example of translational research. The same will take place in the field of microbiomic research.

These new technologies do not diminish the role of physicians, but expand and strengthen their diagnostic, discriminant, and predictive abilities. Although immersed in a universe of computer clouds composed of an enormous amount of data, physicians cannot surrender their humanistic vocation and lose contact with patients as persons. Humanization, or better still, medical humanities, and super technology are in reality two sides of the same coin.

We must go beyond the dualism of the simplicity of protocols and the medicine of complexity. The protocols can certainly be of use in daily practice, but they must not limit the physician's task, which is to take into account the uniqueness of each patient. Protocols are made to be surpassed when necessary: the choices made must be motivated and the clinical line of reasoning on every single patient must be demonstrated. Treatment is an eminently individual issue.

The institutions we have created must provide the right to health care to all, but this does not mean that the response to treatment is the same for all, nor can it be the same for all as dictated by law. We are quite different one from the other: even identical twins are not identical, not even at birth, and their diversity increases more and more as time goes by. The interindividual biological variability that emerges from the -omics technologies is a certainty that we can no longer neglect.

If only some of what I have said here has reached the reader, I consider myself gratified. In any case, the road to the future has been traced and goes in the direction of personalized medicine tailored to the needs of every single patient and capable of providing effective prevention for each individual. To do this we may have to throw some protocols into the wastepaper basket and switch on our brains once again: we must listen to patients and make available to them the formidable new technologies that I have spoken of in this book. Are we ready for this evolution? Are we ready for this revolution?

Bibliography

Books

Alario, F.X., 2013. Essere intelligenti è una malattia? Tutto quello che vorremmo sapere sul cervello. Einaudi, Turin.

Anninos, P., Rossi, M., Pham, T.D., Falugi, C., Bussing, A., Koukkou, M. (Eds.), 2010. Recent advances in clinical medicine. Proceedings of the International Conference on Medical Pharmacology (Pharmacology '10); Proceedings of the International Conference on Medical Histology and Embryology (Histem '10), University of Cambridge, UK, February 23–25, 2010. WSEAS Press, Athens.

Aranda, J.V., Fanos, V., van den Anker, J.N. (Eds.), 2012. Perinatal Pharmacology: Individualized Neonatal Therapy. Hygeia Press, Quartu Sant'Elena.

Avery, G.B., MacDonald, M.G., Seshia, M.M.K., Mullett, M.D. (Eds.), 2005. Avery's Neonatology: Pathophysiology & Management of the Newborn, sixth ed. Lippincott Williams & Wilkins, Philadelphia.

Barabàsi, A.-L., 2011. Lampi. La trama nascosta che guida la nostra vita. Einaudi, Turin.

Bartocci, C., 2014. Dimostrare l'impossibile. Raffaello Cortina, Milan.

Bergson, H., 1998. Le due fonti della morale e della religione. Laterza, Rome-Bari.

Bobbio, M., 2010. Il malato immaginato. I rischi di una medicina senza limiti. Einaudi, Turin.

Boncinelli, E., 1999. Il cervello, la mente e l'anima. Le straordinarie scoperte sull'intelligenza umana. Arnoldo Mondadori, Milan.

Boncinelli, E., 2014. Genetica e guarigione. Einaudi, Turin.

Brillat-Savarin, J.A., 1826. Physiologie du gout, ou méditations de gastronomie trascendante. A. Sautelet et Cie libraires, Paris.

Burgio, G.R., Martini, A., 2000. Pediatria pratica. Orientamenti diagnostici e terapeutici, 2a edizione. Edizioni Medico-Scientifiche, Turin.

Cadeddu, A., D'Aloja, E., Faa, G., Fanos, V., Granese, A., Rutelli, P., 2007. Qualità e sanità: un dialogo per l'umanizzazione. Filosofia, pedagogia, medicina e psicologia. Franco Angeli, Milan.

Cataldi, L., Simeoni, U., Fanos, V., 1996. Neonatal Nephrology in Progress. Agorà, Lecce.

Ceruti, M., 2014. La fine dell'onniscienza. Ed. Studium, Rome.

Charon, R., 2006. Narrative Medicine, Honoring the Stories of Illness. Oxford University Press, Oxford.

Corridori, M., Fanos, V., Farnetani, I., 2007a. Nascere nella storia, 2a edizione. Mondadori Electa, Milan.

Corridori, M., Fanos, V., Farnetani, I., 2007b. Birth through the Ages. Mondadori Electa, Milan.

Cosmacini, G., 2000. Il mestiere di medico. Storia di una professione. Raffaello Cortina, Milan, p. XI.

Cosmacini, G., 2005. L'arte lunga. Storia della medicina dall'antichità ad oggi. Laterza, Rome-Bari.

Cosmacini, G., 2007. La religiosità della medicina. Dall'antichità a oggi. Laterza, Rome-Bari.

Cosmacini, G., 2009. Prima lezione di medicina. Laterza, Rome-Bari.

Metabolomics and Microbiomics. http://dx.doi.org/10.1016/B978-0-12-805305-8.00015-7

Dawkins, R., 1976. The Selfish Gene. Oxford University Press, Oxford.
Delhanty, P.J.D., van der Lely, A.J. (Eds.), 2014. How Gut and Brain Control Metabolism. (Frontiers of Hormone Research), vol. 42, Karger, Basel, pp. 83–92.
Edelman, G.M., 2005. Wider Than the Sky: The Phenomenal Gift of Consciousness. Yale University Press, New Haven, CT.
Faa, G., Fanos, V., 2014. Kidney Development in Renal Pathology. Humana Press, New York.
Faa, G., Fanos, V., Van Eyken, P. (Eds.), 2014. Perinatal Pathology. The Role of the Clinical Pathological Dialogue in Problem Solving. Hygeia Press, Quartu Sant'Elena.
Facchinetti, F., 2014. Esperienza clinica con mioinositolo in ginecologia e ostetricia. Minerva Medica, Turin.
Fanos, V., Yurdakök, M. (Eds.), 2010. Children of the Mother Goddess. History of Mediterranean Neonates. Hygeia Press, Quartu Sant'Elena.
Fanos, V., Corridori, M., Cataldi, L. (Eds.), 2003. Pueri, puerorum, pueris. Miti, storia e credenze sui bambini attraverso i secoli. Agorà, Lecce.
Fanos, V., Chevalier, R.L., Faa, G., Cataldi, L. (Eds.), 2011. Developmental Nephrology: From Embryology to Metabolomics. Hygeia Press, Quartu Sant'Elena.
Gaburro, D., 1997. Pediatria. Generale e specialistica. Idelson-Gnocchi, Naples.
Gershon, M.D., 2013. Il secondo cervello. De Agostini, Novara.
Giarelli, G., Venneri, E., 2009. Sociologia della salute e della medicina. Manuale per le professioni mediche, sanitarie. Franco Angeli, Milan.
Greenspan, S., 1997. The Growth of the Mind and the Endangered Origins of Intelligence. Da Capo Press, Cambridge, MA.
Hartley, J., 1999. Uses of Television. Routledge, London.
Hoffmann, P.M., 2014. Gli ingranaggi di Dio. Dal caos molecolare alla vita. Bollati Boringhieri, Turin.
Howard, J., 2003. Darwin. Il Mulino, Bologna.
Ippocrate, 1994. Aforismi e Giuramento. Newton Compton, Rome.
Kochhar, S., Martin, F.P., 2015. Metabonomics and Gut Microbiota in Nutrition and Disease. Humana Press, Springer Verlag, London.
Kumalè, C., 2007. Il mondo a tavola. Precetti, riti e tabù. Einaudi, Turin.
Lagercrantz, H., 2010. Il cervello del bambino. Come si forma la mente. Giunti Editore, Florence.
Lanza, R. (Ed.), 2009. Essentials of Stem Cell Biology, 2nd ed. Elsevier, Oxford.
Lipton, B.H., 2006. La biologia delle credenze. Come il pensiero influenza il DNA e ogni cellula. Macro Edizioni, Cesena.
Medawar, P.B., Medawar, J.S., 1983. Aristotle to Zoos: A Philosophical Dictionary of Biology. Harvard University Press, Cambridge.
Metchnikoff, E., 1907. The Prolongation of Life. Putman and Sons, New York.
Mortari, L., 2015. Filosofia della cura. Raffaello Cortina, Milan.
Naisbitt, J., Naisbitt, N., Philips, D., 2001. High Tech High Touch: Technology and Our Accelerated Search for Meaning. Nicholas Brealey Publishing, London.
Natterson-Horowitz, B., Bowers, K., 2012. Zoobiquity. What Animals Teach us About Health and the Science of Healing. Alfred Knopf, New York.
Nilsen, M., 2011. Reinventing Discovery: The New Era of Networked Science. Princeton University Press, Princeton, NJ.
Oliver, J., 1968. Nephrons and Kidney: A Quantitative Study of Developmental and Evolutionary Mammalian Renal Architectonics. Harper and Row, New York.
Perlmutter, D., 2015. La dieta intelligente. Perché grano, carboidrati e zuccheri minacciano il nostro cervello. Mondadori, Milan.
Perniola, M., 2004. Contro la comunicazione. Einaudi, Turin.

Pistoi, S., 2012. Il DNA incontra Facebook. Viaggio nel supermercato della genetica. Marsilio, Venice.

Postman, N., Weingartner, C., 2009. Teaching as a Subversive Activity. Dell Publishing, New York.

Rigotti, F., 2004. La filosofia in cucina. Il Mulino, Bologna.

Rotilio, G., 2012. Il migratore onnivoro. Storia e geografia della nutrizione umana. Carocci, Rome.

Schrödinger, E., 1944. What is Life? Cambridge University Press, Cambridge.

Seung, S., 2013. Il connettoma. La nuova geografia della mente. Codice Edizioni, Turin.

Spector, T., 2013. Uguali ma diversi. Quello che i nostri geni non controllano. Bollati Boringhieri, Turin.

Stampolidis, N.C., Tassoulas, Y. (Eds.), 2014. Hygieia. Health, Illness, Treatment from Homer to Galen. Museum of Cycladic Art—Hellenic Ministry of Culture and Sports, Athens, p. 346.

Stokes, D.E., 1997. Pasteur's Quadrant—Basic Science and Technological Innovation. Brookings Institution Press, Washington, DC.

Swaab, D., 2014. We are Our Brains. From the Womb to the Alzheimer's. Penguin Books, London.

Taleb, N.N., 2013. Antifragile. Prosperare nel disordine. Il Saggiatore, Milan.

Tamburini, M., Santosuosso, A., 1999. Malati a rischio. Implicazioni etiche, legali e psicosociali dei test genetici in oncologia. Elsevier-Masson, Milan.

Tomasko, R.M., 2006. Bigger Isn't Always Better: The New Mindset for Real Business Growth. AMACOM, New York.

UNICEF, Curto, C., Fiasco, M. (Eds.), 2014. 25 anni di progressi per l'infanzia e l'adolescenza. UNICEF, Rome.

Venter, J.C., 2014. Il disegno della vita. Dalla mappa del genoma alla biologia digitale. Rizzoli, Milan.

WHO, March of Dimes, PMNCH, Save the Children, 2012. In: Howson, C.P., Kinney, M.V., Lawn, J.E. (Eds.), Born Too Soon: The Global Action Report on Preterm Birth. World Health Organization, Geneva.

Wrangham, R., 2011. L'intelligenza del fuoco. L'invenzione della cottura e l'evoluzione dell'uomo. Bollati Boringhieri, Turin.

Yuste, R., Church, G.M., Lein, E., et al., 2014. Il secolo del cervello. Dai neuroni alla coscienza. Le Scienze, Rome.

Zonza, M. (Ed.), 2013. Medical humanities in infermieristica neonatale. Hygeia Press, Quartu Sant'Elena.

Book chapters

Bai, G., 1990. "Progresso", Grande Dizionario Enciclopedico, IV Edizione. UTET, Turin.

Bulfamante, G., Avagliano, L., 2008. L'indagine anatomo-patologica sulla morte fetale. In: Baronciani, D., Bulfamante, G., Facchinetti, F. (Eds.), La natimortalità: audit clinico e miglioramento della pratica assistenziale. Il Pensiero Scientifico, Rome, pp. 63–100.

Fanos, V., 2008. La comunicazione in Pediatria. In: Cadeddu, A., D'Aloja, E., Faa, G., Fanos, V., Granese, A., Rutelli, P. (Eds.), Qualità e sanità: un dialogo per l'umanizzazione. Franco Angeli, Milan.

Franks, A., 1991. Nanotechnology. In: Gardner, J.W., Hingle, H.T. (Eds.), From Instrumentation to Nanotechnology. Gordon and Breach Science Publishers, Amsterdam.

Menichella, M. (Ed.), 2006. Intervista a Piero Angela. Professione divulgatore. SciBooks Edizioni, Pisa.

Nijland, M.J., Nathanielsz, P.W., 2009. Developmental programming of the kidney. In: Newnham, J.P., Ross, M.G. (Eds.), Early Life Origins of Human Health and Disease. Karger, Basel.

Scientific articles

Aagaard, K., Ma, J., Antony, K.M., Ganu, R., Petrosino, J., Versalovic, J., 2014. The placenta harbors a unique microbiome. Sci. Transl. Med. 6 (237), 1–11.

AAP Section on breastfeeding, 2012. Breastfeeding and the use of human milk. Pediatrics 129 (3), e827–e841.

Abbott, A., 2015. Neuroscience: the brain, interrupted. Nature 518 (7537), 24–26.

Ahn, A.C., Tewari, M., Poon, C.S., Phillips, R.S., 2006. The limits of reductionism in medicine: could systems biology offer an alternative? PLoS Med. 3 (6), e208.

Aiello, L.C., Wells, C.K., 2002. Energetics and the evolution of the genus homo. Annu. Rev. Anthropol. 31, 323–338.

Aiello, L.C., Wheeler, P., 1995. The expensive-tissue hypothesis. The brain and the difensive system in human and primate evolution. Annu. Rev. Anthropol. 36, 199–231.

Altman, D.G., 1994. The scandal of poor medical research. BMJ 308, 283–284.

Arroyo, R., Martín, V., Maldonado, A., Jiménez, E., Fernández, L., Rodríguez, J.M., 2010. Treatment of infectious mastitis during lactation: antibiotics versus oral administration of lactobacilli isolated from breast milk. Clin. Infect. Dis. 50 (12), 1551–1558.

Arumugam, M., Raes, J., Pelletier, E., Le Paslier, D., Yamada, T., Mende, D.R., Fernandes, G.R., Tap, J., Bruls, T., Batto, J.M., Bertalan, M., Borruel, N., Casellas, F., Fernandez, L., Gautier, L., Hansen, T., Hattori, M., Hayashi, T., Kleerebezem, M., Kurokawa, K., Leclerc, M., Levenez, F., Manichanh, C., Nielsen, H.B., Nielsen, T., Pons, N., Poulain, J., Qin, J., Sicheritz-Ponten, T., Tims, S., Torrents, D., Ugarte, E., Zoetendal, E.G., Wang, J., Guarner, F., Pedersen, O., de Vos, W.M., Brunak, S., Doré, J., Meta HIT Consortium, Antolín, M., Artiguenave, F., Blottiere, H.M., Almeida, M., Brechot, C., Cara, C., Chervaux, C., Cultrone, A., Delorme, C., Denariaz, G., Dervyn, R., Foerstner, K.U., Friss, C., van de Guchte, M., Guedon, E., Haimet, F., Huber, W., van Hylckama-Vlieg, J., Jamet, A., Juste, C., Kaci, G., Knol, J., Lakhdari, O., Layec, S., Le Roux, K., Maguin, E., Mérieux, A., Melo Minardi, R., M'rini, C., Muller, J., Oozeer, R., Parkhill, J., Renault, P., Rescigno, M., Sanchez, N., Sunagawa, S., Torrejon, A., Turner, K., Vandemeulebrouck, G., Varela, E., Winogradsky, Y., Zeller, G., Weissenbach, J., Ehrlich, S.D., Bork, P., 2011. Enterotypes of the human gut microbiome. Nature 473 (7346), 174–180.

Ashwell, M., 2000. Obituary: Elsie Widdowson (1906–2000). Nature 406 (6798), 844.

Atzori, L., Xanthos, T., Barberini, L., Antonucci, R., Murgia, F., Lussu, M., Aroni, F., Varsami, M., Papalois, A., Lai, A., D'Aloja, E., Iacovidou, N., Fanos, V., 2010. A metabolomic approach in an experimental model of hypoxia-reoxygenation in newborn piglets: urine predicts outcome. J. Matern. Fetal Neonatal Med. 23 (Suppl. 3), 134–137.

Atzori, L., Mussap, M., Noto, A., Barberini, L., Puddu, M., Irmesi, R., Murgia, F., Lussu, M., Fanos, V., 2011. Clinical metabolomics and urinary NGAL for the early prediction of chronic kidney disease in healthy adults born ELBW. J. Matern. Fetal Neonatal Med. 24, 41–44.

Azevedo, F.A., Carvalho, L.R., Grinberg, L.T., Farfel, J.M., Ferretti, R.E., Leite, R.E., Jacob Filho, W., Lent, R., Herculano-Houzel, S., 2009. Equal numbers of neuronal and non-neuronal cells make the human brain an isometrically scaled-up primate brain. J. Comp. Neurol. 513 (5), 532–541.

Bäckhed, F., Roswall, J., Peng, Y., Feng, Q., Jia, H., Kovatcheva-Datchary, P., Li, Y., Xia, Y., Xie, H., Zhong, H., Khan, M.T., Zhang, J., Li, J., Xiao, L., Al-Aama, J., Zhang, D., Lee,

Y.S., Kotowska, D., Colding, C., Tremaroli, V., Yin, Y., Bergman, S., Xu, X., Madsen, L., Kristiansen, K., Dahlgren, J., Jun, W., 2015. Dynamics and stabilization of the human gut microbiome during the first year of life. Cell Host Microbe 17 (5), 690–703.

Bailey, L.C., Forrest, C.B., Zhang, P., Richards, T.M., Livshits, A., DeRusso, P.A., 2014. Association of antibiotics in infancy with early childhood obesity. JAMA Pediatr. 168 (11), 1063–1069.

Barabási, A.-L., Gulbahce, N., Loscalzo, J., 2011. Network medicine: a network-based approach to human disease. Nat. Rev. Gen. 12 (1), 56–68.

Bardin, C., Piuze, G., Papageorgiou, A., 2004. Outcome at 5 years of age of SGA and AGA infants born less than 28 weeks of gestation. Semin. Perinatol. 28 (4), 288–294.

Barker, D.J.P., 1990. The fetal and infant origins of adult disease. BMJ 301 (6761), 1111.

Barker, D.J.P., 1995. Fetal origin of coronary heart disease. BMJ 311, 171–174.

Barton, R.A., Capellini, I., 2011. Maternal investment, life histories, and the costs of brain growth in mammals. Proc. Natl. Acad. Sci. USA 108 (15), 6169–6174.

Basile, C., Libutti, P., Teutonico, A., Lomonte, C., 2010. Tossine uremiche: il caso dei "protein-bound compounds". G. Ital. Nefrol. 27 (5), 498–507.

Bassareo, P.P., Fanos, V., Barbanti, C., Mercuro, G., 2013. Prematurity at birth and increased cardiovascular risk: is a metabolomic approach the right solution? J. Pediatr. Neonatal Individual. Med. 2 (1), 28–34.

Bhutta, Z.A., Darmstadt, G.L., 2014. A role for science investments in advancing newborn health. Sci. Transl. Med. 6 (253), 253cm8.

Bode, L., McGuire, M., Rodriguez, J.M., Geddes, D.T., Hassiotou, F., Hartmann, P.E., McGuire, M.K., 2014. It's alive: microbes and cells in human milk and their potential benefits to mother and infant. Adv. Nutr. 5 (5), 571–573.

Boekelheide, K., Blumberg, B., Chapin, R.E., Cote, I., Graziano, J.H., Janesick, A., Lane, R., Lillycrop, K., Myatt, L., States, J.C., Thayer, K.A., Waalkes, M.P., Rogers, J.M., 2012. Predicting later-life outcomes of early life exposure. Environ. Health Perspect. 120 (10), 1353–1361.

Boyle, A., Reddy, U.M., 2012. Epidemiology of cesarean delivery: the scope of the problem. Semin. Perinatol. 36 (5), 308–314.

Brandt, L.J., 2012. Fecal transplantation for the treatment of Clostridium difficile infection. Gastroenterol. Hepatol. 8, 191–194.

Buccoliero, L., 2010. E-Health 2.0. Tecnologie per il patient empowerment. Mondo Digitale 4, 3–17.

Buonocore, G., Mussap, M., Fanos, V., 2013. Proteomics and metabolomics: can they solve some mysteries of the newborn? J. Matern. Fetal Neonatal Med. 26 (Suppl. 2), 7–8.

Burgio, E., Migliore, L., 2015. Towards a systemic paradigm in carcinogenesis: linking epigenetics and genetics. Mol. Biol. Rep. 42 (4), 777–790.

Burrell, S.A., Exley, C., 2010. There is (still) too much aluminium in infant formulas. BMC Pediatr. 10, 63.

Caboni, P., Meloni, A., Lussu, M., Carta, E., Barberini, L., Noto, A., Deiana, S.F., Mereu, R., Ragusa, A., Paoletti, A.M., Melis, G.B., Fanos, V., Atzori, L., 2014. Urinary metabolomics of pregnant women at term: a combined GC/MS and NMR approach. J. Matern. Fetal Neonatal Med. 27 (Suppl. 2), 4–12.

Cabrera-Rubio, R., Collado, M.C., Laitinen, K., Salminen, S., Isolauri, E., Mira, A., 2012. The human milk microbiome changes over lactation and is shaped by maternal weight and mode of delivery. Am. J. Clin. Nutr. 96 (3), 544–551.

Cantonwine, D.E., Ferguson, K.K., Mukherjee, B., McElrath, T.F., Meeker, J.D., 2015. Urinary bisphenol A levels during pregnancy and risk of preterm birth. Environ. Health Perspect. 123 (9), 895–901.

Cantor, G.H., 2010. Metabolomics and mechanisms: sometimes the fisher catches a big fish. Toxicol. Sci. 118 (2), 321–323.

Caramia, G., Atzei, A., Fanos, V., 2008. Probiotics and the skin. Clin. Dermatol. 26, 4–11.

Casazza, K., Fontaine, K.R., Astrup, A., Birch, L.L., Brown, A.W., Bohan Brown, M.M., Durant, N., Dutton, G., Foster, E.M., Heymsfield, S.B., McIver, K., Mehta, T., Menachemi, N., Newby, P.K., Pate, R., Rolls, B.J., Sen, B., Smith, Jr., D.L., Thomas, D.M., Allison, D.B., 2013. Myths, presumptions, and facts about obesity. N. Engl. J. Med. 368 (5), 446–454.

Castagnola, M., Uda, F., Noto, A., Fanos, V., Faa, G., 2014. The triple-I (interactive, intersectorial, interdisciplinary) approach to validate "omics" investigations on body fluids and tissues in perinatal medicine. J. Matern. Fetal Neonatal Med. 27 (Suppl. 2), 58–60.

Cavaiuolo, C., Casani, A., Di Manso, G., Orfeo, L., 2015. Effect of Mozart music on heel prick pain in preterm infants: a pilot randomized controlled trial. J. Pediatr. Neonatal Individual. Med. 4 (1), e040109.

Cesare Marincola, F., Noto, A., Caboni, P., Reali, A., Barberini, L., Lussu, M., Murgia, F., Santoru, M.L., Atzori, L., Fanos, V., 2012. A metabolomic study of preterm human and formula milk by high resolution NMR and GC/MS analysis: preliminary results. J. Matern. Fetal Neonatal. Med. 25 (Suppl. 5), 62–67.

Cesare Marincola, F., Dessì, A., Corbu, S., Reali, A., Fanos, V., 2015. Clinical impact of human breast milk metabolomics. Clin. Chim. Acta. 451, 103–106.

Chalmers, I., Bracken, M.B., Djulbegovic, B., Garattini, S., Grant, J., Gülmezoglu, A.M., Howells, D.W., Ioannidis, J.P., Oliver, S., 2014. How to increase value and reduce waste when research priorities are set. Lancet 383 (9912), 156–165.

Christakis, N.A., Fowler, J.H., 2007. The spread of obesity in a large social network over 32 years. N. Engl. J. Med. 357 (4), 370–379.

Claus, S.P., Ellero, S.L., Berger, B., Krause, L., Bruttin, A., Molina, J., Paris, A., Want, E.J., de Waziers, I., Cloarec, O., Richards, S.E., Wang, Y., Dumas, M.E., Ross, A., Rezzi, S., Kochhar, S., Van Bladeren, P., Lindon, J.C., Holmes, E., Nicholson, J.K., 2011. Colonization-induced host-gut microbial metabolic interaction. MBio 2 (2), e00271–10.

Clubb, Jr., F.J., Bishop, S.P., 1984. Formation of binucleated myocardial cells in the neonatal rat. An index for growth hypertrophy. Lab Invest. 50 (5), 571–577.

Collins, F.S., Varmus, H., 2015. A new initiative on precision medicine. N. Engl. J. Med. 372 (9), 793–795.

Cregan, M.D., Fan, Y., Appelbee, A., Brown, M.L., Klopcic, B., Koppen, J., Mitoulas, L.R., Piper, K.M., Choolani, M.A., Chong, Y.S., Hartmann, P.E., 2007. Identification of nestin-positive putative mammary stem cells in human breastmilk. Cell Tissue Res. 329 (1), 129–136.

Curtis, D.J., Sood, A., Phillips, T.J., Leinster, V.H., Nishiguchi, A., Coyle, C., Lacharme-Lora, L., Beaumont, O., Kemp, H., Goodall, R., Cornes, L., Giugliano, M., Barone, R.A., Matsusaki, M., Akashi, M., Tanaka, H.Y., Kano, M., McGarvey, J., Halemani, N.D., Simon, K., Keehan, R., Ind, W., Masters, T., Grant, S., Athwal, S., Collett, G., Tannetta, D., Sargent, I.L., Scull-Brown, E., Liu, X., Aquilina, K., Cohen, N., Lane, J.D., Thoresen, M., Hanley, J., Randall, A., Case, C.P., 2014. Secretions from placenta, after hypoxia/reoxygenation, can damage developing neurones of brain under experimental conditions. Exp. Neurol. 261, 386–395.

D'Aloja, E., Floris, L., Muller, M., Birocchi, F., Fanos, V., Paribello, F., Demontis, R., 2010. Shared decision-making in neonatology: an utopia or an attainable goal? J. Matern. Fetal Neonatal Med. 23 (Suppl. 3), 56–58.

de Snoo, K., 1937. Das trinkende Kind im Uterus. Mschr. Geburt. Gynak. 105, 88–97.

Dekaban, A.S., 1978. Changes in brain weights during the span of human life: relation of brain weights to body heights and body weights. Ann. Neurol. 4 (4), 345–356.

Dessì, A., Atzori, L., Noto, A., Visser, G.H., Gazzolo, D., Zanardo, V., Barberini, L., Puddu, M., Ottonello, G., Atzei, A., De Magistris, A., Lussu, M., Murgia, F., Fanos, V., 2011. Metabolomics in newborns with intrauterine growth retardation (IUGR): urine reveals markers of metabolic syndrome. J. Matern. Fetal Neonatal Med. 24 (Suppl. 2), 35–39.

Dessì, A., Marincola, F.C., Masili, A., Gazzolo, D., Fanos, V., 2014. Clinical metabolomics and nutrition: the new frontier in neonatology and pediatrics. Biomed. Res. Int. 2014, 981219.

Dessì, A., Pravettoni, C., Cesare Marincola, F., Schirru, A., Fanos, V., 2015. The biomarkers of fetal growth in intrauterine growth retardation and large for gestational age cases: from adipocytokines to a metabolomic all-in-one tool. Expert Rev. Proteomics 12 (3), 309–316.

Dominguez-Bello, M.G., Costello, E.K., Contreras, M., Magris, M., Hidalgo, G., Fierer, N., Knight, R., 2010. Delivery mode shapes the acquisition and structure of the initial microbiota across multiple body habitats in newborns. Proc. Natl. Acad. Sci. USA 107 (26), 11971–11975.

Donzelli, G., 2014. The change of paradigm in perinatal sciences: the role of Narrative Medicine and Medical Humanities. J. Pediatr. Neonatal Individual. Med. 3 (2), e030236.

Donzelli, G., 2015. Medical humanities and narrative medicine in perinatal care. J. Matern. Fetal Neonatal Med. 28 (1), 1–2.

Dosa, D.M., 2007. A day in the life of Oscar the Cat. N. Engl. J. Med. 357 (4), 328–329.

Doyle, L.W., Anderson, P., 2010. Adult outcomes of extremely preterm infants. Pediatrics 126, 342–351.

Dunn, W.B., Broadhurst, D.I., Atherton, H.J., Goodacre, R., Griffin, J.L., 2011. Systems level studies of mammalian metabolomes: the roles of mass spectrometry and nuclear magnetic resonance spectroscopy. Chem. Soc. Rev. 40 (1), 387–426.

Duque-Guimarães, D.E., Ozanne, S.E., 2013. Nutritional programming of insulin resistance: causes and consequences. Trends Endocrinol. Metab. 24 (10), 525–535.

Englund-Ögge, L., Brantsæter, A.L., Sengpiel, V., Haugen, M., Birgisdottir, B.E., Myhre, R., Meltzer, H.M., Jacobsson, B., 2014. Maternal dietary patterns and preterm delivery: results from large prospective cohort study. BMJ 348, g1446.

Enos, W.F., Holmes, R.H., Beyer, J., 1953. Coronary disease among United States soldiers killed in action in Korea; preliminary report. J. Am. Med. Assoc. 152 (12), 1090–1093.

Faa, G., Gerosa, C., Fanni, D., Nemolato, S., Locci, A., Cabras, T., Marinelli, V., Puddu, M., Zaffanello, M., Monga, G., Fanos, V., 2010. Marked interindividual variability in renal maturation of preterm infants: lessons from autopsy. J. Matern. Fetal Neonatal Med. 23 (Suppl. 3), 129–133.

Faa, G., Gerosa, C., Fanni, D., Monga, G., Zaffanello, M., Van Eyken, P., Fanos, V., 2012a. Morphogenesis and molecular mechanisms involved in human kidney development. J. Cell Physiol. 227 (3), 1257–1268.

Faa, G., Gerosa, C., Fanni, D., Nemolato, S., Di Felice, E., Van Eyken, P., Monga, G., Iacovidou, N., Fanos, V., 2012b. The role of immunohistochemistry in the study of the newborn kidney. J. Matern. Fetal Neonatal Med. 25 (Suppl. 4), 135–138.

Faa, G., Nemolato, S., Cabras, T., Fanni, D., Gerosa, C., Fanari, M., Locci, A., Fanos, V., Messana, I., Castagnola, M., 2012c. Thymosin β4 expression reveals intriguing similarities between fetal and cancer cells. Ann. N.Y. Acad. Sci. 1269, 53–60.

Faa, G., Gerosa, C., Fanni, D., Nemolato, S., van Eyken, P., Fanos, V., 2013. Factors influencing the development of a personal tailored microbiota in the neonate are well known, with particular emphasis on antibiotic therapy. J. Matern. Fetal Neonatal Med. 26 (Suppl. 2), 35–43.

Faa, G., Marcialis, M.A., Ravarino, A., Piras, M., Pintus, M.C., Fanos, V., 2014a. Fetal programming of the human brain: is there a link with insurgence of neurodegenerative disorders in adulthood? Curr. Med. Chem. 21 (33), 3854–3876.

Faa, A., Ambu, R., Faa, G., Fanos, V., 2014b. Perinatal heart programming: long-term consequences. Curr. Med. Chem. 21 (27), 3165–3172.

Faa, G., Sanna, A., Gerosa, C., Fanni, D., Puddu, M., Ottonello, G., Van Eyken, P., Fanos, V., 2015. Renal physiological regenerative medicine to prevent chronic renal failure: Should we start at birth? Clin. Chim. Acta. 444, 156–162.

Fanni, D., Fanos, V., Monga, G., Gerosa, C., Locci, A., Nemolato, S., Van Eyken, P., Faa, G., 2011a. Expression of WT1 during normal human kidney development. J. Matern. Fetal Neonatal Med. 24 (Suppl. 2), 44–47.

Fanni, D., Fanos, V., Monga, G., Gerosa, C., Nemolato, S., Locci, A., Van Eyken, P., Iacovidou, N., Faa, G., 2011b. MUC1 in mesenchymal-to-epithelial transition during human nephrogenesis: changing the fate of renal progenitor/stem cells? J. Matern. Fetal Neonatal Med. 24 (Suppl. 2), 63–66.

Fanni, D., Gerosa, C., Nemolato, S., Mocci, C., Pichiri, G., Coni, P., Congiu, T., Piludu, M., Piras, M., Fraschini, M., Zaffanello, M., Iacovidou, N., Van Eyken, P., Monga, G., Faa, G., Fanos, V., 2012a. “Physiological” renal regenerating medicine in VLBW preterm infants: could a dream come true? J. Matern. Fetal Neonatal Med. 25 (Suppl. 3), 41–48.

Fanni, D., Iacovidou, N., Locci, A., Gerosa, C., Nemolato, S., Van Eyken, P., Monga, G., Mellou, S., Faa, G., Fanos, V., 2012b. MUC1 marks collecting tubules, renal vesicles, comma- and S-shaped bodies in human developing kidney. Eur. J. Histochem. 56 (4), e40.

Fanni, D., Ambu, R., Gerosa, C., Nemolato, S., Iacovidou, N., Van Eyken, P., Fanos, V., Zaffanello, M., Faa, G., 2014a. Aluminum exposure and toxicity in neonates: a practical guide to halt aluminum overload in the prenatal and perinatal periods. World J. Pediatr. 10 (2), 101–107.

Fanni, D., Fanos, V., Gerosa, C., Sanna, A., Van Eyken, P., Cataldi, L., Faa, G., 2014b. Acute kidney injury in the newborn: the role of the perinatal pathologist. J. Pediatr. Neonatal Individual. Med. 3 (2), e030262.

Fanos, V., 2012a. Cells, the tree, medicines and the tailor. Curr. Pharm. Des. 18, 2995.

Fanos, V., 2012b. Pediatric and neonatal individualized Medicine: care and cure for each and everyone. J. Pediatr. Neonatal Individual. Med. 1 (1), 7–10.

Fanos, V., Yurdakok, M., 2010. Personalized neonatal medicine. J. Matern. Fetal Neonatal Med. 23, 4–6.

Fanos, V., Puddu, M., Reali, A., Atzei, A., Zaffanello, M., 2010. Perinatal nutrient restriction reduces nephron endowment increasing renal morbidity in adulthood: a review. Early Hum. Dev. 86 (Suppl. 1), 37–42.

Fanos, V., Antonucci, R., Atzori, L., 2013a. Metabolomics in the developing infant. Curr. Opin. Pediatr. 25 (5), 604–611.

Fanos, V., Atzori, L., Makarenko, K., Melis, G.B., Ferrazzi, E., 2013b. Metabolomics application in maternal-fetal medicine. Biomed. Res. Int. 2013, 720514.

Fanos, V., Fanni, C., Ottonello, G., Noto, A., Dessì, A., Mussap, M., 2013c. Metabolomics in adult and pediatric nephrology. Molecules 18 (5), 4844–4857.

Fanos, V., Van den Anker, J., Noto, A., Mussap, M., Atzori, L., 2013d. Metabolomics in neonatology: fact or fiction? Semin. Fetal Neonatal Med. 18 (1), 3–12.

Fanos, V., Buonocore, G., Mussap, M., 2014a. Neonatomics and childomics: the right route to the future. J. Matern. Fetal Neonatal Med. 27 (Suppl. 2), 1–3.

Fanos, V., Mussap, M., Faa, G., Papageorgiou, A., 2014b. The next ten years in neonatology: new directions in research. J. Pediatr. Neonatal Individual. Med. 3 (2), e030239.

Fanos, V., Noto, A., Caboni, P., Pintus, M.C., Liori, B., Dessì, A., Mussap, M., 2014c. Urine metabolomic profiling in neonatal nephrology. Clin. Biochem. 47 (9), 708–710.

Fanos, V., Noto, A., Xanthos, T., Lussu, M., Murgia, F., Barberini, L., Finco, G., d'Aloja, E., Papalois, A., Iacovidou, N., Atzori, L., 2014d. Metabolomics network characterization of resuscitation after normocapnic hypoxia in a newborn piglet model supports the hypothesis that room air is better. Biomed. Res. Int. 2014, 731620.

Fanos, V., Pintus, M.C., Lussu, M., Atzori, L., Noto, A., Stronati, M., Guimaraes, H., Marcialis, M.A., Rocha, G., Moretti, C., Papoff, P., Lacerenza, S., Puddu, S., Giuffrè, M., Serraino, F., Mussap, M., Corsello, G., 2014e. Urinary metabolomics of bronchopulmonary dysplasia (BPD): preliminary data at birth suggest it is a congenital disease. J. Matern. Fetal Neonatal Med. 27 (Suppl. 2), 39–45.

Fanos, V., Loddo, C., Puddu, M., Gerosa, C., Fanni, D., Ottonello, G., Faa, G., 2015. From ureteric bud to the first glomeruli: genes, mediators, kidney alterations. Int. Urol. Nephrol. 47 (1), 109–116.

Farnetani, I., Fanos, V., 2014. David Barker: the revolution that anticipates existence. J. Pediatr. Neonatal Individual. Med. 3 (1), e030111.

Fernández, L., Langa, S., Martín, V., Maldonado, A., Jiménez, E., Martín, R., Rodríguez, J.M., 2013. The human milk microbiota: origin and potential roles in health and disease. Pharmacol. Res. 69 (1), 1–10.

Feynman, R.P., 1960. There's plenty of room at the bottom. An invitation to enter a new field of physics. Eng. Sci. 23 (5), 22–36.

Fox, T.P., Godavitarne, C., 2012. What really causes necrotising enterocolitis? ISRN Gastroenterol. 2012, 628317.

Gierman, H.J., Fortney, K., Roach, J.C., Coles, N.S., Li, H., Glusman, G., Markov, G.J., Smith, J.D., Hood, L., Coles, L.S., Kim, S.K., 2014. Whole-genome sequencing of the world's oldest people. PLoS One 9 (11), e112430.

Goddeeris, J.H., Saigal, S., Boyle, M.H., Paneth, N., Streiner, D.L., Stoskopf, B., 2010. Economic outcomes in young adulthood for extremely low birth weight survivors. Pediatrics 126 (5), e1102–e1108.

Gohir, W., Ratcliffe, E.M., Sloboda, D.M., 2015. Of the bugs that shape us: maternal obesity, the gut microbiome, and long-term disease risk. Pediatr. Res. 77 (1-2), 196–204.

Griffiths, S.K., Brown, Jr., W.S., Gerhardt, K.J., Abrams, R.M., Morris, R.J., 1994. The perception of speech sounds recorded within the uterus of a pregnant sheep. J. Acoust. Soc. Am. 96 (4), 2055–2063.

Guerrera, G., 2015. Neonatal and pediatric healthcare worldwide: a report from UNICEF. Clin. Chim. Acta. 451, 4–8.

Gustafsson, M., Nestor, C.E., Zhang, H., Barabási, A.L., Baranzini, S., Brunak, S., Chung, K.F., Federoff, H.J., Gavin, A.C., Meehan, R.R., Picotti, P., Pujana MÀ, Rajewsky, N., Smith, K.G., Sterk, P.J., Villoslada, P., Benson, M., 2014. Modules, networks and systems medicine for understanding disease and aiding diagnosis. Genome Med. 6 (10), 82.

Hales, C.N., Barker, D.J.P., 1992. Type 2 (non-insulin-dependent) diabetes mellitus: the thrifty phenotype hypothesis. Diabetologia 35, 595–601.

Hales, C.N., Barker, D.J.P., 2001. The thrifty phenotype hypothesis. Br. Med. Bull. 60, 5–20.

Heazell, A.E., Brown, M., Worton, S.A., Dunn, W.B., 2011. The effects of oxygen on normal and pre-eclamptic placental tissue—insights from metabolomics. Placenta 32 (Suppl. 2), S119–S124.

Hochberg, Z., 2011. Developmental plasticity in child growth and maturation. Front. Endocrinol. (Lausanne) 2, 41.

Hochberg, Z., Feil, R., Constancia, M., Fraga, M., Junien, C., Carel, J.C., Boileau, P., Le Bouc, Y., Deal, C.L., Lillycrop, K., Scharfmann, R., Sheppard, A., Skinner, M., Szyf, M., Waterland, R.A., Waxman, D.J., Whitelaw, E., Ong, K., Albertsson-Wikland, K., 2011. Child health, developmental plasticity, and epigenetic programming. Endocr. Rev. 32 (2), 159–224.

Hoffman, A., Spengler, D., 2014. DNA memories of early social life. Neuroscience 264, 64–75.

Holland, D., Chang, L., Ernst, T.M., Curran, M., Buchthal, S.D., Alicata, D., Skranes, J., Johansen, H., Hernandez, A., Yamakawa, R., Kuperman, J.M., Dale, A.M., 2014. Structural growth trajectories and rates of change in the first 3 months of infant brain development. JAMA Neurol. 71 (10), 1266–1274.

Hood, L., Friend, S.H., 2011. Predictive, personalized, preventive, participatory (P4) cancer medicine. Nat. Rev. Clin. Oncol. 8 (3), 184–187.

Hosseini, S.M., Talaei-Khozani, T., Sani, M., Owrangi, B., 2014. Differentiation of human breast-milk stem cells to neural stem cells and neurons. Neurol. Res. Int. 2014, 807896.

Ide, M., Papapanou, P.N., 2013. Epidemiology of association between maternal periodontal disease and adverse pregnancy outcomes – systematic review. J. Clin. Periodontol. 40 (Suppl. 14), S181–S194.

Jiménez, E., Fernández, L., Maldonado, A., Martín, R., Olivares, M., Xaus, J., Rodríguez, J.M., 2008. Oral administration of Lactobacillus strains isolated from breast milk as an alternative for the treatment of infectious mastitis during lactation. Appl. Environ. Microbiol. 74 (15), 4650–4655.

Jonsen, A.R., 2012. Morality in the valley of the moon: the origins of the ethics of neonatal intensive care. Theor. Med. Bioeth. 33 (1), 65–74.

Kang, D.W., Park, J.G., Ilhan, Z.E., Wallstrom, G., Labaer, J., Adams, J.B., Krajmalnik-Brown, R., 2013. Reduced incidence of Prevotella and other fermenters in intestinal microflora of autistic children. PLoS One 8 (7), e68322.

Karachaliou, M., Georgiou, V., Roumeliotaki, T., Chalkiadaki, G., Daraki, V., Koinaki, S., Dermitzaki, E., Sarri, K., Vassilaki, M., Kogevinas, M., Oken, E., Chatzi, L., 2015. Association of trimester-specific gestational weight gain with fetal growth, offspring obesity, and cardiometabolic traits in early childhood. Am. J. Obstet. Gynecol. 212 (4), 502.e1–502.e14.

Kemp, M.W., 2014. Preterm birth, intrauterine infection, and fetal inflammation. Front. Immunol. 5, 574.

Kenneth, J.G., Abrams, R.M., 2000. Fetal exposures to sound and vibroacoustic stimulation. J. Perinatol. 20 (Suppl. 8), S21–S30.

Kisilevsky, B.S., Hains, S.M., Brown, C.A., Lee, C.T., Cowperthwaite, B., Stutzman, S.S., Swansburg, M.L., Lee, K., Xie, X., Huang, H., Ye, H.H., Zhang, K., Wang, Z., 2009. Fetal sensitivity to properties of maternal speech and language. Infant Behav. Dev. 32 (1), 59–71.

Koletzko, B., Brands, B., Chourdakis, M., Cramer, S., Grote, V., Hellmuth, C., Kirchberg, F., Prell, C., Rzehak, P., Uhl, O., Weber, M., 2014. The Power of Programming and the Early-Nutrition project: opportunities for health promotion by nutrition during the first thousand days of life and beyond. Ann. Nutr. Metab. 64 (3-4), 187–196.

Kotze, H.L., Armitage, E.G., Sharkey, K.J., Allwood, J.W., Dunn, W.B., Williams, K.J., Goodacre, R., 2013. A novel untargeted metabolomics correlation-based network analysis incorporating human metabolic reconstructions. BMC Syst. Biol. 7, 107.

Kuzawa, C.W., Hallal, P.C., Adair, L., Bhargava, S.K., Fall, C.H., Lee, N., Norris, S.A., Osmond, C., Ramirez-Zea, M., Sachdev, H.S., Stein, A.D., Victora, C.G., 2012. Birth weight, postnatal weight gain, and adult body composition in five low and middle income countries. Am. J. Hum. Biol. 24 (1), 5–13.

Lackritz, E.M., Wilson, C.B., Guttmacher, A.E., Howse, J.L., Engmann, C.M., Rubens, C.E., Mason, E.M., Muglia, L.J., Gravett, M.G., Goldenberg, R.L., Murray, J.C., Spong, C.Y., Simpson, J.L., Preterm Birth Research Priority Setting Group, 2013. A solution pathway for preterm birth. Lancet Glob. Health. 1 (6), e328–e330.

Lichtman, J.W., Pfister, H., Shavit, N., 2014. The big data challenges of connectomics. Nat. Neurosci. 17 (11), 1448–1454.

Liley, A.W., 1972. The foetus as a personality. Aust. NZ. J. Psychiatry 6 (2), 99–105.

Lindsay, K.L., Walsh, C.A., Brennan, L., McAuliffe, F.M., 2013. Probiotics in pregnancy and maternal outcomes: a systematic review. J. Matern. Fetal Neonatal Med. 26 (8), 772–778.

Lok, C., 2005. Metabolomics. A new diagnostic tool could mean spotting diseases earlier and more easily. MIT Technol. Rev. Special issue: 10 Emerging Technologies. 108(5):46-47.

Luan, H., Meng, N., Liu, P., Feng, Q., Lin, S., Fu, J., Davidson, R., Chen, X., Rao, W., Chen, F., Jiang, H., Xu, X., Cai, Z., Wang, J., 2014. Pregnancy-induced metabolic phenotype variations in maternal plasma. J. Proteome Res. 13 (3), 1527–1536.

Luke, B., Mamelle, N., Keith, L., Munoz, F., Minogue, J., Papiernik, E., Johnson, T.R., Research Committee of the Association of Women's Health, Obstetric, and Neonatal Nurses, 1995. The association between occupational factors and preterm birth: a United States nurses' study. Am. J. Obstet. Gynecol. 173 (3 Pt 1), 849–862.

Maldonado, J., Cañabate, F., Sempere, L., Vela, F., Sánchez, A.R., Narbona, E., López-Huertas, E., Geerlings, A., Valero, A.D., Olivares, M., Lara-Villoslada, F., 2012. Human milk probiotic Lactobacillus fermentum CECT5716 reduces the incidence of gastrointestinal and upper respiratory tract infections in infants. J. Pediatr. Gastroenterol. Nutr. 54 (1), 55–61.

Mamelle, N., Laumon, B., Lazar, P., 1984. Prematurity and occupational activity during pregnancy. Am. J. Epidemiol. 119 (3), 309–322.

Marois, R., Ivanoff, J., 2005. Capacity limits of information processing in the brain. Trends Cogn. Sci. 9 (6), 296–305.

Martignoni, G., 2007. Uno stile per "pensare in altra luce". Riv. Med. Hum. 1 (1), 25–30.

Martin, F.P., Rezzi, S., Peré-Trepat, E., Kamlage, B., Collino, S., Leibold, E., Kastler, J., Rein, D., Fay, L.B., Kochhar, S., 2009. Metabolic effects of dark chocolate consumption on energy, gut microbiota, and stress-related metabolism in free-living subjects. J. Proteome Res. 8 (12), 5568–5579.

Martin, F.P., Collino, S., Rezzi, S., Kochhar, S., 2012a. Metabolomic applications to decipher gut microbial metabolic influence in health and disease. Front. Physiol. 3, 113.

Martin, F.P., Montoliu, I., Nagy, K., Moco, S., Collino, S., Guy, P., Redeuil, K., Scherer, M., Rezzi, S., Kochhar, S., 2012b. Specific dietary preferences are linked to differing gut microbial metabolic activity in response to dark chocolate intake. J. Proteome Res. 11 (12), 6252–6263.

Martín, R., Olivares, M., Marín, M.L., Fernández, L., Xaus, J., Rodríguez, J.M., 2005. Probiotic potential of 3 lactobacilli strains isolated from breast milk. J. Hum. Lact. 21, 8–17.

Martins-de-Souza, D., 2014. Proteomics, metabolomics, and protein interactomics in the characterization of the molecular features of major depressive disorder. Dialog. Clin. Neurosci. 16 (1), 63–73.

Matamoros, S., Gras-Leguen, C., Le Vacon, F., Potel, G., de La Cochetiere, M.-F., 2013. Development of intestinal microbiota in infants and its impact on health. Trends Microbiol. 21 (4), 167–173.

Mayhew, T.M., Gregson, C., Pharaoh, A., Fagan, D.G., 1998. Numbers of nuclei in different tissue compartments of fetal ventricular myocardium from 16 to 35 weeks of gestation. Virchows Arch. 433 (2), 167–172.

McCay, C.M., 1933. Is longevity compatible with optimum growth? Science 77, 410–411.

McDonald, A.D., McDonald, J.C., Armstrong, B., Cherry, N.M., Nolin, A.D., Robert, D., 1988. Prematurity and work in pregnancy. Br. J. Ind. Med. 45 (1), 56–62.

McHardy, I.H., Goudarzi, M., Tong, M., Ruegger, P.M., Schwager, E., Weger, J.R., Graeber, T.G., Sonnenburg, J.L., Horvath, S., Huttenhower, C., McGovern, D.P., Fornace, Jr., A.J., Borneman, J., Braun, J., 2013. Integrative analysis of the microbiome and metabolome of the human intestinal mucosal surface reveals exquisite inter-relationships. Microbiome 1 (1), 17.

Meaney, M.J., 2001. Nature, nurture, and the disunity of knowledge. Ann. NY Acad. Sci. 935, 50–61.

Messerli, F.H., 2012. Chocolate consumption, cognitive function, and nobel laureates. N. Engl. J. Med. 367, 1562–1564.

Metcalfe, N.B., Monaghan, P., 2001. Compensation for a bad start: grow now, pay later? Trends Ecol. Evol. 16 (5), 254–260.

Michaelsen, K.F., 2010. WHO growth standards. Should they be implemented as national standards? J. Pediatr. Gastroenterol. Nutr. 51 (Suppl. 3), S151–S152.

Michel, J.B., Shen, Y.K., Aiden, A.P., Veres, A., Gray, M.K., Google Books Team, Pickett, J.P., Hoiberg, D., Clancy, D., Norvig, P., Orwant, J., Pinker, S., Nowak, M.A., Aiden, E.L., 2011. Quantitative analysis of culture using millions of digitized books. Science 331 (6014), 176–182.

Mocci, C., Congiu, T., Fanni, D., Gerosa, C., Komuta, M., Van Eyken, P., Faa, G., Riva, A., 2014. Scanning electron microscopy in liver biopsy interpretation in children: a mini atlas. J. Pediatr. Neonatal Individual. Med. 3 (2), e030206.

Mönckeberg, J.G., 1915. Uber die Atherosklerose der Kombattanten (each Obdurtionsbefunden). Zentralbl Herz Gefasskrankheiten 7, 10–22.

Moon, C., Panneton Cooper, R., Fifer, W.P., 1993. Two-day-olds prefer their native language. Infant Behav. Dev. 16, 495–500.

Morris, G., Berk, M., 2015. The many roads to mitochondrial dysfunction in neuroimmune and neuropsychiatric disorders. BMC Med. 13, 68.

Muglia, L.J., Katz, M., 2010. The enigma of spontaneous preterm birth. N. Engl. J. Med. 362 (6), 529–535.

Musch, T.I., 2014. Discipuli supra omni praeferuntur: above all, the students come first. Physiologist 57 (1), 1, 48–50.

Mussap, M., 2014. An alternative perspective on how laboratory medicine can contribute to solve the health care crisis: a model to save costs by acquiring excellence in diagnostic systems. Clin. Chim. Acta. 427, 202–204.

Mussap, M., Antonucci, R., Noto, A., Fanos, V., 2013. The role of metabolomics in neonatal and pediatric laboratory medicine. Clin. Chim. Acta. 426, 127–138.

Myhre, R., Brantsæter, A.L., Myking, S., Eggesbø, M., Meltzer, H.M., Haugen, M., Jacobsson, B., 2013. Intakes of garlic and dried fruits are associated with lower risk of spontaneous preterm delivery. J. Nutr. 143 (7), 1100–1108.

Naeye, R.L., Burt, L.S., Wright, D.L., Blanc, W.A., Tatter, D., 1971. Neonatal mortality, the male disadvantage. Pediatrics 48, 902–906.

Nasidze, I., Li, J., Quinque, D., Tang, K., Stoneking, M., 2009. Global diversity in the human salivary microbiome. Genome Res. 19 (4), 636–643.

Nichols, B.L., 2003. Icie Macy and Elsie Widdowson: pioneers of child nutrition and growth. J. Nutr. 133 (11), 3690–3692.

Nicholson, J.K., Lindon, J.C., 2008. Systems biology: metabonomics. Nature 455 (7216), 1054–1056.

Noto, A., Fanos, V., Barberini, L., Grapov, D., Fattuoni, C., Zaffanello, M., Casanova, A., Fenu, G., De Giacomo, A., De Angelis, M., Moretti, C., Papoff, P., Di Tonno, R.,

Francavilla, R., 2014. The urinary metabolomics profile of an Italian autistic children population and their unaffected siblings. J. Matern. Fetal Neonatal Med. 27 (Suppl. 2), 46–52.

Nurse, P., 2003. The great ideas of biology. Clin. Med. 3 (6), 560–568.

Ogbonnaya, E.S., Clarke, G., Shanahan, F., Dinan, T.G., Cryan, J.F., O'Leary, O.F., 2015. Adult hippocampal neurogenesis is regulated by the microbiome. Biol. Psychiatry 78 (4), e7–e9.

Owyang, C., Wu, G.D., 2014. The gut microbiome in health and disease. Gastroenterology 146, 1433–1436.

Palomo, A.B.A., Lucas, M., Dilley, R.J., McLenachan, S., Chen, F.K., Requena, J., Sal, M.F., Lucas, A., Alvarez, I., Jaraquemada, D., Edel, M.J., 2014. The power and the promise of cell reprogramming: personalized autologous body organ and cell transplantation. J. Clin. Med. 3, 373–387.

Papageorgiou, A., Pelausa, E., 2014. Management and outcome of extremely low birth weight infants. J. Pediatr. Neonatal Individual. Med. 3 (2), e030209.

Parkinson, J.R., Hyde, M.J., Gale, C., Santhakumaran, S., Modi, N., 2013. Preterm birth and the metabolic syndrome in adult life: a systematic review and meta-analysis. Pediatrics 131 (4), e1240–e1263.

Patel, C.J., Yang, T., Hu, Z., Wen, Q., Sung, J., El-Sayed, Y.Y., Cohen, H., Gould, J., Stevenson, D.K., Shaw, G.M., Ling, X.B., Butte, A.J., March of Dimes Prematurity Research Center at Stanford University School of Medicine, 2014. Investigation of maternal environmental exposures in association with self-reported preterm birth. Reprod. Toxicol. 45, 1–7.

Patel, R.M., Kandefer, S., Walsh, M.C., Bell, E.F., Carlo, W.A., Laptook, A.R., Sánchez, P.J., Shankaran, S., Van Meurs, K.P., Ball, M.B., Hale, E.C., Newman, N.S., Das, A., Higgins, R.D., Stoll, B.J., Eunice Kennedy Shriver National Institute of Child Health and Human Development Neonatal Research Network, 2015. Causes and timing of death in extremely premature infants from 2000 through 2011. N. Engl. J. Med. 372 (4), 331–340.

Patki, S., Kadam, S., Chandra, V., Bhonde, R., 2010. Human breast milk is a rich source of multipotent mesenchymal stem cells. Hum. Cell 23 (2), 35–40.

Pellegrino, E., 1979. The most humane science: some notes on liberal education in medicine and the university (The Sixth Sanger Lecture). Bull. Med. Coll. Virginia 67 (4), 11–39.

Perpiñá Tordera, M., 2009. Complexity in asthma: inflammation and scale-free networks. Arch. Bronconeumol. 45 (9), 459–465, (Article in Spanish).

Perrone, S., Tataranno, L.M., Stazzoni, G., Ramenghi, L., Buonocore, G., 2013. Brain susceptibility to oxidative stress in the perinatal period. J. Matern. Fetal Neonatal Med. 28 (Suppl. 1), 2291–2295.

Petrof, E.O., Khoruts, A., 2014. From stool transplants to next-generation microbiota therapeutics. Gastroenterology 146 (6), 1573–1582.

Petrovsky, N., 2001. Towards a unified model of neuroendocrine-immune interaction. Immunol. Cell Biol. 79 (4), 350–357.

Piras, M., Fanos, V., Ravarino, A., Marcialis, M.A., Vinci, L., Pintus, M.C., Faa, G., 2014. Fetal programming of Parkinson's and Alzheimer's diseases: the role of epigenetic factors. J. Pediatr. Neonatal Individual. Med. 3 (2), e030270.

Porrello, E.R., Widdop, R.E., Delbridge, L.M., 2008. Early origins of cardiac hypertrophy: does cardiomyocyte attrition programme for pathological 'catch-up' growth of the heart? Clin. Exp. Pharmacol. Physiol. 35 (11), 1358–1364.

Poste, G., 2011. Bring on the biomarkers. Nature 469 (7329), 156–157.

Prince, L.A., Antony, A.M., Chu, D.M., Aagaard, K.M., 2014. The microbiome, parturition, and timing of birth: more questions than answers. J. Reprod. Immunol. 104-105, 12–19.

Proal, A.D., Albert, P.J., Marshall, T.G., 2014. Inflammatory disease and the human microbiome. Discov. Med. 17 (95), 257–265.

Puddu, M., Fanos, V., 2012. Developmental programming of auditory learning. J. Pediatr. Neonatal Individual. Med. 1 (1), 59–66.

Radford, E.J., Ito, M., Shi, H., Corish, J.A., Yamazawa, K., Isganaitis, E., Seisenberger, S., Hore, T.A., Reik, W., Erkek, S., Peters, A.H., Patti, M.E., Ferguson-Smith, A.C., 2014. In utero undernourishment perturbs the adult sperm methylome and intergenerational metabolism. Science 345 (6198), 1255903.

Ramezani, A., Raj, D.S., 2014. The gut microbiome, kidney disease, and targeted interventions. J. Am. Soc. Nephrol. 25 (4), 657–670.

Rando, O.J., Simmons, R.A., 2015. I'm eating for two: parental dietary effects on offspring metabolism. Cell 161 (1), 93–105.

Reali, A., Greco, F., Fanaro, S., Atzei, A., Puddu, M., Moi, M., Fanos, V., 2010. Fortification of maternal milk for very low birth weight (VLBW) pre-term neonates. Early Hum. Dev. 86 (Suppl. 1), 33–36.

Reali, A., Greco, F., Marongiu, G., Deidda, F., Atzeni, S., Campus, R., Dessì, A., Fanos, V., 2015. Individualized fortification of breast milk in 41 Extremely Low Birth Weight (ELBW) preterm infants. Clin. Chim. Acta. 451, 107–110.

Rees, S., Harding, R., Walker, D., 2011. The biological basis of injury and neuroprotection in the fetal and neonatal brain. Int. J. Dev. Neurosci. 29, 551–563.

Rescigno, M., Urbano, M., Valzasina, B., Francolín, M., Rotta, G., Bonasio, R., Granucci, F., Kraehenbuhl, J.P., Ricciardi-Castagnoli, P., 2001. Dendritic cells express tight junction proteins and penetrate gut epithelial monolayers to sample bacteria. Nat. Immunol. 2 (4), 361–367.

Romero, R., Mazaki-Tovi, S., Vaisbuch, E., Kusanovic, J.P., Chaiworapongsa, T., Gomez, R., Nien, J.K., Yoon, B.H., Mazor, M., Luo, J., Banks, D., Ryals, J., Beecher, C., 2010. Metabolomics in premature labor: a novel approach to identify patients at risk for preterm delivery. J. Matern. Fetal Neonatal Med. 23 (12), 1344–1359.

Rosenfeld, R.M., 2010. Passion. Otolaryngol. Head Neck Surg. 143, 737–738.

Sani, M., Hosseini, S.M., Salmannejad, M., Aleahmad, F., Ebrahimi, S., Jahanshahi, S., Talaei-Khozani, T., 2015. Origins of the breast milk-derived cells; an endeavor to find the cell sources. Cell Biol. Int. 39 (5), 611–618.

Sanz, M., Kornman, K., Working group 3 of joint EFP/AAP workshop, 2013. Periodontitis and adverse pregnancy outcomes: consensus report of the Joint EFP/AAP Workshop on Periodontitis and Systemic Diseases. J. Clin. Periodontol. 40 (Suppl. 14), S164–S169.

Schell, L.M., 1981. Environmental noise and human prenatal growth. Am. J. Phys. Anthropol. 56, 63–70.

Schmidt, C., 2015. Mental health: thinking from the gut. Nature 518 (7540), S12–S15.

Serino, M., Chabo, C., Burcelin, R., 2012. Intestinal MicrobiOMICS to define health and disease in human and mice. Curr. Pharm. Biotechnol. 13 (5), 746–758.

Shafquat, A., Joice, R., Simmons, S.L., Huttenhower, C., 2014. Functional and phylogenetic assembly of microbial communities in the human microbiome. Trends Microbiol. 22 (5), 261–266.

Simmons, R., 2011. Epigenetics and maternal nutrition: nature v. nurture. Proc. Nutr. Soc. 70 (1), 73–81.

Singer, P., 1983. Sanctity of life or quality of life? Pediatrics 72 (1), 128–129.

Slyepchenko, A., Carvalho, A.F., Cha, D.S., Kasper, S., McIntyre, R.S., 2014. Gut emotions—mechanisms of action of probiotics as novel therapeutic targets for depression and anxiety disorders. CNS Neurol. Disord. Drug Targets. 13 (10), 1770–1786.

Smith, G.D., 2011. Epidemiology, epigenetics and the 'Gloomy Prospect': embracing randomness in population health research and practice. Int. J. Epidemiol. 40 (3), 537–562.

Solberg, R., Enot, D., Deigner, H.P., Koal, T., Scholl-Bürgi, S., Saugstad, O.D., Keller, M., 2010. Metabolomic analyses of plasma reveals new insights into asphyxia and resuscitation in pigs. PLoS One 5 (3), e9606.

Standley, J.M., 2002. A meta-analysis of the efficacy of music therapy for premature infants. J. Pediatr. Nurs. 17 (2), 107–113.

Stoll, B.J., Hansen, N.I., Bell, E.F., Shankaran, S., Laptook, A.R., Walsh, M.C., Hale, E.C., Newman, N.S., Schibler, K., Carlo, W.A., Kennedy, K.A., Poindexter, B.B., Finer, N.N., Ehrenkranz, R.A., Duara, S., Sánchez, P.J., O'Shea, T.M., Goldberg, R.N., Van Meurs, K.P., Faix, R.G., Phelps, D.L., Frantz, 3rd, I.D., Watterberg, K.L., Saha, S., Das, A., Higgins, R.D., Eunice Kennedy Shriver National Institute of Child Health and Human Development Neonatal Research Network, 2010. Neonatal outcomes of extremely preterm infants from the NICHD Neonatal Research Network. Pediatrics 126 (3), 443–456.

Sulek, K., Han, T.L., Villas-Boas, S.G., Wishart, D.S., Soh, S.E., Kwek, K., Gluckman, P.D., Chong, Y.S., Kenny, L.C., Baker, P.N., 2014. Hair metabolomics: identification of fetal compromise provides proof of concept for biomarker discovery. Theranostics 4 (9), 953–959.

Tandoi, F., Agosti, M., 2012. Gender differences: are there differences even in Pediatrics and Neonatology? J. Pediatr. Neonatal Individual. Med. 1 (1), 43–48.

Tarazi, C., Agostoni, C., Kim, K.S., 2014. The placental microbiome and pediatric research. Pediatr. Res. 76 (3), 218–219.

Tojo, R., Suárez, A., Clemente, M.G., de los Reyes-Gavilán, C.G., Margolles, A., Gueimonde, M., Ruas-Madiedo, P., 2014. Intestinal microbiota in health and disease: role of bifidobacteria in gut homeostasis. World J. Gastroenterol. 20 (41), 15163–15176.

Twigger, A.J., Hodgetts, S., Filgueira, L., Hartmann, P.E., Hassiotou, F., 2013. From breast milk to brains: the potential of stem cells in human milk. J. Hum. Lact. 29 (2), 136–139.

Upadhyay, K., Pourcyrous, M., Dhanireddy, R., Talati, A.J., 2015. Outcomes of neonates with birth weight ≤500 g: a 20-year experience. J. Perinatol. 35 (9), 768–772.

Waddington, C.H., 1942. The epigenotype. Endeavor 1, 18–20.

Wang, Y., Kasper, L.H., 2014. The role of microbiome in central nervous system disorders. Brain Behav. Immun. 38, 1–12.

Wang-Sattler, R., Yu, Z., Herder, C., Messias, A.C., Floegel, A., He, Y., Heim, K., Campillos, M., Holzapfel, C., Thorand, B., Grallert, H., Xu, T., Bader, E., Huth, C., Mittelstrass, K., Döring, A., Meisinger, C., Gieger, C., Prehn, C., Roemisch-Margl, W., Carstensen, M., Xie, L., Yamanaka-Okumura, H., Xing, G., Ceglarek, U., Thiery, J., Giani, G., Lickert, H., Lin, X., Li, Y., Boeing, H., Joost, H.G., de Angelis, M.H., Rathmann, W., Suhre, K., Prokisch, H., Peters, A., Meitinger, T., Roden, M., Wichmann, H.E., Pischon, T., Adamski, J., Illig, T., 2012. Novel biomarkers for pre-diabetes identified by metabolomics. Mol. Syst. Biol. 8, 615.

Widdowson, E.M., 1980. Adventures in nutrition over half a century. Proc. Nutr. Soc. 39, 239–306.

Widdowson, E.M., 1985. Development of the digestive system: comparative animal studies. Am. J. Clin. Nutr. 41 (Suppl. 2), 384–390.

Wikoff, W.R., Anfora, A.T., Liu, J., Schultz, P.G., Lesley, S.A., Peters, E.C., Siuzdak, G., 2009. Metabolomics analysis reveals large effects of gut microflora on mammalian blood metabolites. Proc. Natl. Acad. Sci. USA 106 (10), 3698–3703.

Wilcock, A., Begley, P., Stevens, A., Whatmore, A., Victor, S., 2015. The metabolomics of necrotising enterocolitis in preterm babies: an exploratory study. J. Matern. Fetal Neonatal Med. 29 (5), 758–762.

Wishart, D.S., 2015. Is cancer a genetic disease or a metabolic disease? EBioMedicine 2(6): 478–479.

Zonza, M., 2012. Narrative based medicine and neonatology: an interpretative approach. J. Pediatr. Neonatal Individual. Med. 1 (1), 49–52.

Popular science articles

[Author not indicated], 2014. By their fridges ye shall know them. Natl. Geogr. 223 (12), 64–65.

Cavazza, C., Peluso, G., 2010. Mappe del dopo Genoma. Il Sole 24 ore. Jun 27, 39.

Di Todaro, F., 2012. Scopri chi sei con le molecole. Inizia l'era della metabolomica: che cosa cambia, dalla diagnosi ai farmaci. La Stampa. Available from: http://www.lastampa.it/2012/10/31/scienza/scopri-chi-sei-con-le-molecole-KYNnqLtSmjNh8mxDUjT4uN/pagina.html

Donzelli, G., 2014. Il cambiamento. Oltre i protocolli. Molto Meyer 2 (1), 21.

Francis, D., Kaufer, D., 2011. Beyond nature vs. nurture. The Scientist. Available from: http://www.the-scientist.com/?articles.view/articleNo/31233/title/Beyond-Nature-vs--Nurture/

Grady, D., 2014. The mysterious tree of a newborn's life. The push to understand the placenta. New York Times Jul 14.

Hall, S.S., 2013. On beyond 100. Natl. Geogr. (5), 28–49.

Interlandi, J., 2013. Oltre le difese del cervello. Le Scienze (8), 38–43.

Kluger, J., 2014. A preemie revolution. Cutting-edge medicine and dedicated caregivers are helping the tiniest babies survive—and thrive. Time Apr 28, 24–31.

McGreevy, K.S., Donzelli, G., 2014. Un racconto allitterativo sul futuro della pediatria. Molto Meyer 2 (4), 4–6.

Meli, E., 2008. La firma del nostro organismo. Corriere della Sera Feb 3, 54.

Miller, P., 2012. A thing… or two about twins. Natl. Geogr. 221 (1), 38–65.

Murphy P.A., 2010. How the First Nine Months Shape the Rest of Your Life. Time. Available from: http://time.com/84145/how-the-first-nine-months-shape-the-rest-of-your-life/

Nestler, J.N., 2012. Il codice epigenetico della mente. Le Scienze (2), 65–71.

Piattelli Palmarini, M., 2014. Domani i microbi ci salveranno. Corriere della Sera Jan 27, 27.

Pievani, T., 2015. Ripensare Darwin? Le Scienze (561), 42–47.

Pope, V., 2014. Un ricchezza condivisa. Natl. Geogr. (12), 20–31.

Quammen, D., 2004. Was Darwin wrong? No. The evidence of evolution is overhelming. Natl. Geogr. (11), 4–35.

Rothstein, E., 2004. The brain? It's a jungle in there. The New York Times. Available from: http://www.nytimes.com/2004/03/27/books/the-brain-it-s-a-jungle-in-there.html

Shaw, J., 2011. Fathoming metabolism. The study of metabolites does an end run around genomics to provide telling clues to your future health. Harvard Mag. (3), 27–31.

Shaw, J., 2014. Why "Big Data" is a big deal. Information science promises to change the world. Harvard Mag. (2), 30–35, 74–75.

Skinner, M.K., 2014. Un nuovo tipo di eredità. Le Scienze (10), 54–61.

Stipp, D., 2012. Una nuova strada verso la longevità. Le Scienze (3), 41–47.

Taino, D., 2012. Elogio dell'antifragilità. Corriere della Sera. Available from: http://lettura.corriere.it/debates/elogio-dellantifragilita/

The Economist, 2012. The human microbiome: me, myself, us. The Economist Aug 18.

Wolfe, N., 2013. Come è piccolo il mondo. Sono invisibili. Sono ovunque. E comandano loro. Natl. Geogr. Italia. 31 (1), 106–117.

Wollheim, R., 1978. Adrian Stokes, critic, painter, poet. Times Literary Supplement Feb 17, 207–209.

Zimmer, C., 2011. 100 Trillion connections: new efforts probe and map the brain's detailed architecture. Sci. Am. (1).

Zimmer, C., 2014. Secrets of the brain. Natl. Geogr. (2), 32–57.

Exhibitions and meetings

11th International Workshop on Developmental Nephrology, 2010. Genetic Programming and the Kidney: Progenitors, Signaling and Morphogenesis in Health and Disease. New Paltz, NY. August 24–27.

Balle di scienza, 2014. Storie di errori prima e dopo Galileo. Palazzo Blu, Pisa. March 22–June 29.

Techorevolution, 2013. The age of converging technologies. Temporary exhibition. Cosmocaixa, Barcelona. January.

Online material

Aggression in Children: unraveling gene-environment interplay to inform Treatment and InterventiON strategies—ACTION, 2015. Available from: http://www.action-euproject.eu

Coordinating Action Systems Medicine—CASyM, 2015. Available from: https://www.casym.eu

EFCNI, 2015. A healthy pregnancy All that matters now. Available from: http://www.efcni.org/fileadmin/Daten/Web/Pregnancy/Healthy_Pregnancy/EFCNI_Brochure_A_healthy_pregnancy_print_final.pdf

Ioannes Paulus PP. II, 2015. Encyclical Letter Evangelium Vitae. Available from: http://w2.vatican.va/content/john-paul-ii/en/encyclicals/documents/hf_jp-ii_enc_25031995_evangelium-vitae.html

Istat, 2014. Gravidanza, parto e allattamento al seno. Anno 2013. Available from: http://www.istat.it/it/files/2014/12/gravidanza.pdf?title=Gravidanza%2C+parto+e+allattamento+al+seno+-+09%2Fdic%2F2014+-+Testo+integrale.pdf

Journal of Pediatric and Neonatal Individualized Medicine. Available from: http://www.jpnim.com

March of Dimes Foundation, 2012. Why at least 39 weeks is best for your baby. Available from: http://www.marchofdimes.org/pregnancy/why-at-least-39-weeks-is-best-for-your-baby.aspx

Michel, J.B., Shen, Y.K., Aiden, A.P., Veres, A., Gray, M.K., Google Books Team, Pickett, J.P., Hoiberg, D., Clancy, D., Norvig, P., Orwant, J., Pinker, S., Nowak, M.A., Aiden, E.L., 2011. Supporting Online Material for: quantitative analysis of culture using millions of digitized books. Science 331 (6014), 176–182, Available from: http://www.sciencemag.org/content/suppl/2010/12/16/science.1199644.DC1/Michel.SOM.revision.2.pdf

Mordor Intelligence, 2015. Global Metabolomics Market Growth, Trends & Forecasts (2014–2020). Description. Available from: http://www.mordorintelligence.com/industry-reports/global-metabolomics-market-industry

Onder, L., 2014. La comunicazione in sanità. Il ruolo e la responsabilità dei media. Lectio doctoralis Luciano Onder in occasione del conferimento della Laurea Magistrale Honoris Causa in Medicina e Chirurgia. Università degli Studi di Parma, Facoltà di Medicina e Chirurgia. Parma, 31 marzo 2014. Available from: http://www.unipr.it/sites/default/files/allegatiparagrafo/31-03-2014/tesi_onder_medicina_chirurgia_stampa.pdf

Sacred Congregation for the Doctrine of the Faith, 2015. Declaration on euthanasia. Available from: http://www.vatican.va/roman_curia/congregations/cfaith/documents/rc_con_cfaith_doc_19800505_euthanasia_en.html

Seung, S., 2010. I am my connectome [Video file]. Available from: http://www.ted.com/talks/sebastian_seung

Films

Gattaca, directed by Andrew Niccol, 1997.

Live Fast, Die Young, directed by Paul Henreid, 1958.

Multiplicity, directed by Harold Ramis, 1996.

The Lord of the Rings, trilogy directed by Peter Jackson, 2001–2003.
Vertigo, directed by Alfred Hitchcock, 1958.

Works of art

Cinema Redux™: creating a visual fingerprint of an entire movie, project by Brendan Dawes, 2004, at the MoMA of New York.

Pala Montefeltro (or Brera Altarpiece), by Piero della Francesca, 1472 circa, at the Brera Picture Gallery in Milan.

Refrigerators, by Mark Menjivar, photographic project begun in 2007, of the series "You Are What You Eat".

Stele found in Athens and now in the British Museum, cited in: Smith AG. A Catalogue of Sculpture in the Department of Greek and Roman Antiquities, British Museum, I. London: 1892. For the inscription: IG II^2, 4513; CIG 606; Hicks EL. The Collection of Ancient Greek Inscriptions in the British Museum, I. Oxford: 1874, pp. 141–142, n. 81.

The Baptism, by Jan Steen, 1663, at the Gemäldegalerie of Berlin.

The Peasant Wedding, by Breugel the Elder, about 1568, at the Kunsthistorisches Museum of Vienna.

Musical works

Symphony no. 40 in Sol minor, K 550, by Wolfgang Amadeus Mozart, 1778.

The Firebird, ballet with music by Igor Stravinsky, 1910.

Philosophical and literary works

Aristotele, 2000. Metafisica (Italian translation edited by Reale G, Radice R). Bompiani, Milan, libro H, 1045 a 9–10.

de Saint-Exupéry, A., 2000. Il Piccolo Principe (Le Petit Prince, Italian translation). Bompiani, Milan.

Eliot, T.S., 2009. Four Quartets. Faber & Faber, London.

Eliot, T.S., 2010. Cori da "La Rocca" (Choruses from "The Rock", Italian translation). Rizzoli, Milan.

Montaigne, 1997. Saggi, libro II, chap. XII (Essais, Italian translation). Mondadori, Milan, p. 604.

Neruda, P., 2004. Per nascere son nato (Para nacer he nacido, Italian translation). Guanda, Milan.

Plato, 1969. Plato in Twelve Volumes, vols. 5 & 6 (English translation by P. Shorey). Harvard University Press, Cambridge, MA; William Heinemann Ltd., London. Republic, book III, 401d.

Süskind, P., 2015. Perfume. The story of a murderer (English translation by J.E. Woods). Penguin UK, London.

Author Index

P

R

S

T

V

W

Z

Subject Index